International and Interdisciplinary Workshop on Novel Phenomena in Integrated Complex Sciences 2010

From Non-living to Living Systems

Journal of Physics: Conference Series Volume 344

Kyoto, Japan
11-14 October 2010

Editors:

Masatoshi Murase
Kazuo Nishimura
Kazuyoshi Yoshimura
Hiroto Ohta

ISBN: 978-1-62276-690-1
ISSN: 1742-6588

Printed from e-media with permission by:

Curran Associates, Inc.
57 Morehouse Lane
Red Hook, NY 12571

Some format issues inherent in the e-media version may also appear in this print version.

Printed by Curran Associates, Inc. (2013)

For permission requests, please contact the Institute of Physics
at the address below.

Institute of Physics
Dirac House, Temple Back
Bristol BS1 6BE UK

Phone: 44 1 17 929 7481
Fax: 44 1 17 920 0979

techtracking@iop.org

Additional copies of this publication are available from:

Curran Associates, Inc.
57 Morehouse Lane
Red Hook, NY 12571 USA
Phone: 845-758-0400
Fax: 845-758-2634
Email: curran@proceedings.com
Web: www.proceedings.com

TABLE OF CONTENTS

012001 Evolution of Quantum Criticality in the System $CeNi_9Ge_4$ 1
H Michor, D T Adroja, A D Hillier, M M Koza, S Manalo, C Gold, L Peyker, E -W Scheidt

012002 Magnetic Isotherm of Itinerant Electron Magnets – A New Approach to Itinerant Electron Metamagnetism 9
Y Takahashi

012003 Correlation Between the Superconducting Pairing Symmetry and Magnetic Anisotropy in f-electron Unconventional Superconductors 19
S Kambe, H Sakai, Y Tokunaga, K Kaneko

012004 Thinking Patterns, Brain Activity and Strategy Choice 27
K Nishimura, A Okada, M Inagawa, Y Tobinaga

012005 Movement Regulation of a Sliding Actin Filament in a Reconstruction Motility Assay System 33
I Kunita, S Sakurazawa

012006 The Effect of a 50-Hz Sinusoidal Magnetic Field on Nitric Oxide (NO) Production by Human Umbilical Vein Endothelial Cells (HUVECs) 39
H Miyata, K-I Ishizawa, M Ishido, K Sugawara, M Murase, T Hondou

012007 Transcriptional Effects of 50 Hz Magnetic Fields at 1.2 μT and 100 μT on Human Breast Cancer MCF-7 Cells 48
M Ishido, H Miyata, K-I Ishizawa, M Murase, T Hondou

012008 A Spatiotemporal Structure: Common to Subatomic Systems, Biological Processes, and Economic Cycles 53
K Naitoh

012009 Topological Aspects of the Quantum Spin Nanotube 71
Y Otsuka, T Sakai

012010 Synthesis and Magnetic Properties of Cu Oxides with Low-dimensional Structure 77
M Kato, Y Nogi, S Hongo, K Hirota

012011 Geometric Frustration in the Stella Quadrangula Lattice and Metamagnetism in Fe_3Mo_3N 83
H Nakamura, S Terazawa, T Waki, Y Tabata, M Takigawa

012012 Superconducting Properties and Thermal Expansion of $YbGa_xSi_{2-x}$ 88
N Tsujii, M Imai, H Kitazawa

012013 Optical Conductivity of $Ca_{1-x}Na_xV_2O_4$ (x = 0, 2/3 and 1) 96
H Sakurai, A Irizawa, T Nanba

012015 Effect of Dimensionality Crossover on Magnetovolume Properties of Quasi One-dimensional Weakly Antiferromagnetic Metals 100
N Hatayama, R Konno, Y Takahashi

012016 Volume Effects and Pressure Effects on Ferromagnetic Superconductors 107
R Konno, N Hatayama

012017 Magnetic Susceptibility of η-carbide-type Molybdenum and Tungsten Carbides and Nitrides 113
T Waki, S Terazawa, Y Umemoto, Y Tabata, Y Murase, M Kato, K Hirota, H Nakamura

012018 The Relation between the Electric Conductance of Nanostructure Bridge and Friedel Sum Rule 117
Y Kotani, N Shima, K Makoshi

012019 Magnetic Properties of Itinerant Ferromagnet $La_{12}Fe_{57.5}As_{41}$ 125
B Chen, J Yang, S Takehara, H Ohta, C Michioka, K Yoshimura

012020 Probing Onset of Coherence in $CeIrIn_5$ 129
A C Shockley, A P Dioguardi, N apRoberts-Warren, P Klavins, N J Curro

012021 Understanding of Superconductivity in Elemental Substances by Pseudopotential Radii 133
Y Makino, K Yoshimura

012022 Local Magnetic Inhomogeneities Observed Via ^{75}As NMR in $(Fe_{1-x}Nix)_2As_2$ with $H_0 \perp$ C-axis 139
A P Dioguardi, N apRoberts-Warren, A C Shockley, S L Bud'Ko, N Ni, P C Canfield, N J Curro

012023 Magnetic Properties of the Novel Layered Compounds $RE_3Ru_4Al_{12}$ (RE = La-Nd) 143
W Ge, H Ohta, C Michioka, K Yoshimura

012024 Single Crystal Growth and Magnetic Property of NH_4CuPO_4 - H_2O 148
K Nawa, C Michioka, K Yoshimura, A Matsuo, K Kindo

012025 Magnetic Properties of Layered Compounds LnCoAsO (Ln = Lanthanoids)) with Itinerant-Electron Ferromagnetism 153
H Ohta, C Michioka, K Yoshimura

012026 ^{77}Se-NMR Study of Quasi-one Dimensional Organic Conductor $(TMTSF)_2X$ 160
K Nishiyama, T Mito, Y Kujirai, T Koyama, K Ueda, T Kohara, K Takeuchi, H Akutsu, J Yamada, A Kornilov, V M Pudalov, J S Qualls
012027 Nuclear Quadrupole Resonance in the Heavy Fermion Antiferromagnet $CePt_2In_7$ 165
N apRoberts-Warren, A P Dioguardi, A C Shockley, C H Lin, J Crocker, P Klavins, N J Curro
Author Index

Journal of Physics: Conference Series **344** (2012) 011001 doi:10.1088/1742-6596/344/1/011001

PREFACE

In this workshop recent advancements in experiments and theories were discussed on magnetism and superconductivity, emergent phenomena in biological material, chemical properties and economic problems of non-living and living systems. The aim of the workshop was to discuss old, but also new problems from a multidisciplinary perspective, and to understand the general features behind diversity in condensed matter physics, experimental chemistry and physics in biology and economic science. The workshop was broadly based, and was titled 'International & Interdisciplinary Workshop on Novel Phenomena in Integrated Complex Sciences from Non-living to Living Systems'. However, the primary focus was on magnetism and superconductivity, and NMR research into strongly correlated electrons. The meeting was held as an ICAM workshop, upon official approval in January 2010. Both young scientists and graduate students were invited. We hope that these young scientists had the chance to talk with invited speakers and organizers on their own interests. We thank the participants who contributed through their presentations, discussions and these papers to the advancement of the subject and our understanding. The proceedings are published here in the *Journal of Physics: Conference Series* (UK).

We thank the International Advisory Committee for their advice and guidance:

Evgeny Antipov (Moscow State University, Russia)
Nicholas Curro (University of California, Davis, USA)
Minghu Fang (Zhejiang University, China)
Jurgen Haase (University of Leipzig, Germany)
Takashi Imai (McMaster University, Canada)
Peter Lemmens (TU Braunschweig, Germany)
Herwig Michor (Vienna TU, Austria)
Takamasa Momose (University of British Columbia, Canada)
Raivo Stern (NICPB, Estonia)
Louis Taillefer (University of Sherbrooke, Canada)
Masashi Takigawa (University of Tokyo, Japan)

This workshop was mainly organized by the International Research Unit of Integrated Complex System Science, Kyoto University, and was supported by ICAM (Institute for Complex and Adaptive Matter, USA), Yukawa Institute for Theoretical Physics (Kyoto University), Institute of Economic Research (Kyoto University) and Kyoto University GCOEs (Global Centers Of Excellence: Physics, Chemistry, and Economics). The workshop was also supported by Niki Glass Company Ltd., THAMWAY Corp., TAIYO NIPPON SANSO, and Quantum Design Japan.

The Editors and the Organizing Committee,

Masatoshi Murase (Kyoto University, Japan),
Kazuo Nishimura (Kyoto University, Japan),
Kazuyoshi Yoshimura (Kyoto University, Japan): *Conference Chairman and Chief Editor,*
Hiroto Ohta (Kyoto University, Tokyo University of A&T, Japan): *Conference Secretary*

CONFERENCE PHOTOGRAPH

14 October 2010

Journal of Physics: Conference Series **344** (2012) 011001 doi:10.1088/1742-6596/344/1/011001

CONFERENCE POSTER

International and Interdisciplinary Workshop on

Novel Phenomena in Integrated Complex Sciences : from Non-living to Living Systems

Organizing Committee

Masatoshi Murase
(Kyoto University, Japan): Vice-Chair

Kazuo Nishimura
(Kyoto University, Japan)

Kazuyoshi Yoshimura
(Kyoto University, Japan): Chair

October
11 (mon) - 14 (thu)
2010
Co-op Inn Kyoto
Conference Hall

International Advisory Committee*
& Invited Speakers#

Evgeny Antipov*# (Moscow State University, Russia)
Nicholas Curro*# (University of California Davis, USA)
Minghu Fang*# (Zhejiang University, China)
Satoshi Fujimoto# (Kyoto University, Japan)
Jürgen Haase*# (Universität Leipzig, Germany)
Takashi Imai*# (McMaster University, Canada)
Kenji Ishida# (Kyoto University, Japan)
Masayuki Itoh# (Nagoya University, Japan)
Shinsaku Kambe# (JAEA, Japan)
Peter Lemmens*# (TU Braunschweig, Germany)
Herwig Michior*# (TU Wien, Austria)
Takamasa Momose*# (University of British Columbia, Canada)
Hiroyuki Nakamura# (Kyoto University, Japan)
Yuichi Shimakawa# (Kyoto University, Japan)
Raivo Stern*# (NI of Chem. Phys. & Biophys., Estonia)
Hiroshi Sugiyama# (Kyoto University, Japan)
Louis Taillefer* (Universite de Sherbrooke, Canada)
Yoshinori Takahashi# (University of Hyogo, Japan)
Masashi Takigawa*# (University of Tokyo, Japan)
Hiroaki Ueda# (University of Tokyo, Japan)
Yoichi Yanase# (Niigata University, Japan)

Organized by

The International Research Unit of Integrated Complex System Sciences (Kyoto University),
ICAM (Institute for Complex and Adaptive Matter, USA),
Yukawa Institute for Theoretical Physics (Kyoto University),
Institute of Economic Research (Kyoto University),
and Kyoto University GCOEs (Global Center Of Excellence: Physics, Chemistry, and Economics)

http://kuchem.kyoto-u.ac.jp/kinso/IIW2010/HOME.html

All papers published in this volume of *Journal of Physics: Conference Series* have been peer reviewed through processes administered by the proceedings Editors. Reviews were conducted by expert referees to the professional and scientific standards expected of a proceedings journal published by IOP Publishing.

Evolution of quantum criticality in the system $CeNi_9Ge_4$

H Michor[1], D T Adroja[2], A D Hillier[2], M M Koza[3], S Manalo[1], C Gold[4], L Peyker[4] and E-W Scheidt[4]

[1]Institut für Festkörperphysik, Technische Universität Wien, A-1040 Wien, Austria
[2]ISIS Facility, Rutherford Appleton Laboratory Chilton, Didcot OX11 0QX, UK
[3]Institut LaueLangevin, B.P. 156, F-38042, Grenoble Cedex 9, France
[4] CPM, Institut für Physik, Universität Augsburg, 86159 Augsburg, Germany

E-mail: michor@ifp.tuwien.ac.at

Abstract. The heavy fermion system $CeNi_9Ge_4$ exhibits a paramagnetic ground state with remarkable features such as: a record value of the electronic specific heat coefficient in systems with a paramagnetic ground state, $\gamma = C/T \simeq 5.5$ J/mol K^2 at 80 mK, a temperature-dependent Sommerfeld–Wilson ratio, $R = \chi/\gamma$, below 1 K and an approximate single ion scaling of the $4f$-magnetic specific heat and susceptibility. These features are related to a rather small Kondo energy scale of a few Kelvin in combination with a quasi-quartet crystal field ground state. Tuning the system towards long range magnetic order is accomplished by replacing a few at.% of Ni by Cu or Co. Specific heat, susceptibility and resistivity studies reveal $T_N \sim 0.2$ K for $CeNi_8CuGe_4$ and $T_N \sim 1$ K for $CeNi_8CoGe_4$. To gain insight whether the transition from the paramagnetic NFL state to the magnetically ordered ground state is connected with a heavy fermion quantum critical point we performed specific heat and ac susceptibility studies and utilized the μSR technique and quasi-elastic neutron scattering. [1]

1. Introduction

Some exciting novel electronic phenomena in solids, e.g. high-temperature and heavy-Fermion superconductivity, have been observed when tuning strongly correlated electron systems by substitution doping, pressure or other non-thermal parameters from a symmetry breaking ground state towards a symmetry conserving one, e.g. from long range magnetic order towards a paramagnetic Fermi liquid ground state. Prominent examples in this context are pressure induced, magnetically mediated unconventional superconducting states in antiferromagnetic (AF) Kondo lattice systems $CePd_2Si_2$, $CeIn_3$ [1] and in ferromagnetic UGe_2 [2]. If continuous, such symmetry breaking phase transition at virtually zero temperature gives rise to quantum critical behavior at finite temperature (for review see e.g. [3]) which e.g. manifests in the non-Fermi liquid (NFL) behavior of various heavy Fermion systems [4, 5].

$CeNi_9Ge_4$ is a very interesting paramagnetic Kondo lattice system that exhibits remarkable features such as a record value of the electronic specific heat coefficient with a paramagnetic ground state, $\gamma = C/T \simeq 5.5$ J/mol K^2 at 80 mK [6], which is the highest among the strongly

[1] This workshop was supported in part by the Grant-in-Aid for the Global COE Program *The Next Generation of Physics, Spun from Universality and Emergence* from the Ministry of Education, Culture, Sports, Science and Technology (MEXT) of Japan.

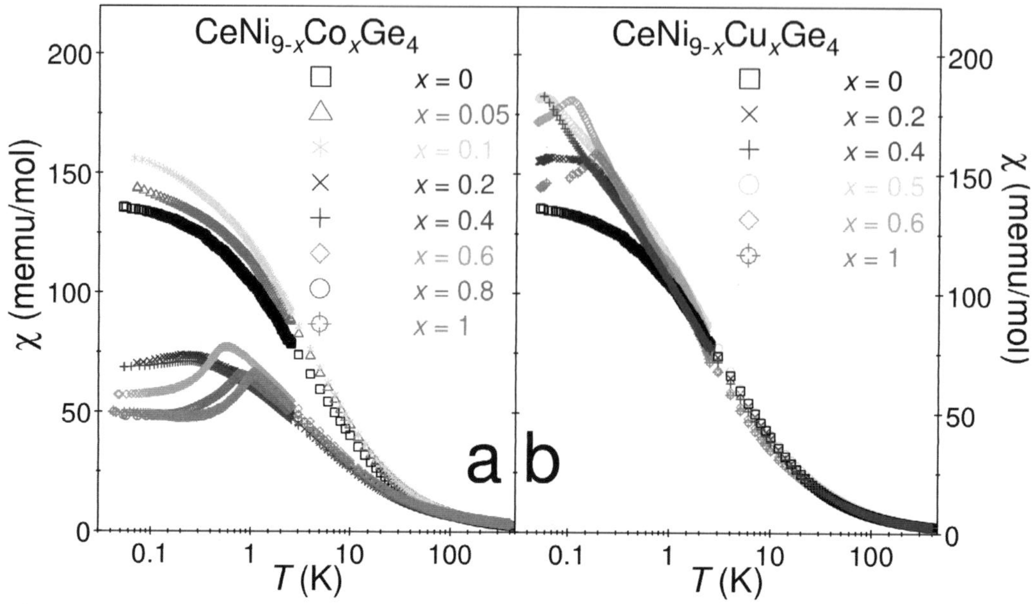

Figure 1. The magnetic susceptibility $\chi(T)$ of $CeNi_{9-x}Co_xGe_4$ (a) and $CeNi_{9-x}Cu_xGe_4$ (b) in semi-logarithmic plots (redrawn from [13, 14]). AFM transitions are evident for $x > 0.5$.

correlated electron systems. $CeNi_9Ge_4$ further exhibits a strongly temperature-dependent Sommerfeld–Wilson ratio, $R = \chi/\gamma$, below 1 K and an approximate scaling of the $4f$-magnetic specific heat and susceptibility for the Ce concentration in a magnetically dilute solid solution $Ce_xLa_{1-x}Ni_9Ge_4$ [7]. Such scaling of the magnetic contributions to the specific heat and susceptibility reveals that the single ion type interactions, such as crystal field (CF) and Kondo effects, are responsible for the physical properties of these compounds. The CF ground state of $CeNi_9Ge_4$ has been determined by means of single crystal susceptibility measurements which were analyzed in terms of the tetragonal crystal field model in combination with a Kondo screening correction by the poor mans scaling approach yielding a CF level scheme with a ground state formed by two doublets ($\Gamma_7^{(1)}$, $\Gamma_7^{(2)}$) split by about 0.5 meV and a well separated Γ_6 doublet with an excitation energy of 11 meV [8]. The low-temperature Kondo energy scale referring to the quasi-quartet ground state which results from the analysis of the macroscopic static susceptibility as well as microscopic neutron studies of the quasi-elastic linewidth is about 0.3 meV [8]. The splitting of the $\Gamma_7^{(1)}$ and $\Gamma_7^{(2)}$ doublets is, thus, of comparable magnitude as the Kondo energy scale and thereby smeared to a quasi-quartet ground state which shows up in the specific heat as roughly $C \sim -T \ln T$, which is a typical behaviour observed in NFL systems at low temperature [6, 7].

Based on the above estimates of CF and Kondo energy scales of $CeNi_9Ge_4$ numerical renormalization group calculations using the $SU(4)$ Anderson impurity model accounted for the basic thermodynamic features of $CeNi_9Ge_4$ reasonably well and demonstrated that the temperature dependent Sommerfeld–Wilson ratio $R(T) = \chi(T)/\gamma(T)$ results from an $SU(2)$ to $SU(4)$ cross-over [9, 10].

Cerium Kondo lattice systems having Kondo temperatures T_K as low as a few Kelvin usually exhibit a magnetically ordered ground state [11], i.e. they are dominated by RKKY inter-site interactions. $CeNi_9Ge_4$ with $T_K \sim 3.5$ K and no magnetic frustration (tetragonal structure with easy direction of magnetisation in c-axis), however, displays a paramagnetic ground state. A factor in favor of a paramagnetic ground state is the effectively quasi-fourfold ground state degeneracy as proposed in the theoretical model of Coleman [12]. In this model above a critical value of the Kondo coupling constant, $(J\rho)_c$, the spin-compensated Kondo-lattice ground state is stable and this value $(J\rho)_c$ is shown to tend to zero as $O(1/N)$, providing new justification of applicability of the Kondo lattice model for rare earth based strongly correlated electron systems [12]. To explore the significance of RKKY interactions in $CeNi_9Ge_4$ and to check for the possibility of quantum criticality in this system, we have studied Ni-site substitutions by Cu and Co, i.e. equivalent to electron and hole doping, respectively. The aim to tune the system towards long range magnetic order is in fact accomplished by replacing a few at.% of Ni by Co as well as by Cu (see figure 1 and references [13, 14] for further details including specific heat and resistivity studies). It is surprising to find magnetic order for both, $CeNi_8CoGe_4$ and $CeNi_8CuGe_4$, because electron and hole doping are expected to drive the Kondo temperature in opposite directions. The latter is in fact supported by the susceptibility data displayed in figure 1 where Co substitution in 1a clearly reduces the susceptibility [14], i.e. indicating an increase of T_K, whereas Cu substitution in 1b tends to increase the low temperature susceptibility, thus, pointing towards a reduction of T_K in the solid solution $CeNi_{9-x}Cu_xGe_4$ [13]. What both solid solutions, $CeNi_{9-x}T_xGe_4$ with T = Co and Cu, show in common is a reduction of the effectively quasi-fourfold degeneracy of the CF ground state of the parent $CeNi_9Ge_4$ compound towards a two-fold degenerate one in both $CeNi_8CoGe_4$ and $CeNi_8CuGe_4$ which is proposed by the analysis of magnetic entropy data [13, 14].

In this paper we present new microscopic studies of magnetic correlations in $CeNi_{9-x}Cu_xGe_4$ by means of quasi-elastic neutron scattering and muon spin relaxation (μSR) experiments. Macroscopic studies of thermodynamic properties such as magnetic susceptibility, specific heat and thermal expansion revealed a quantum critical point in this solid solution series where $CeNi_{8.6}Cu_{0.4}Ge_4$ exhibits quantum critical behavior with χ and $C/T \propto -\ln T$ as well as a strongly composition dependent behavior of the Grüneisen ratio $\Gamma(T) \propto \alpha(T)/C(T)$ at low temperatures where $\Gamma(T \to 0)$ increases by orders of magnitudes right at the critical composition $CeNi_{8.6}Cu_{0.4}Ge_4$ [13].

2. Experimental details

Polycrystalline samples used for neutron scattering and μSR studies namely $CeNi_9Ge_4$, $CeNi_{8.6}Cu_{0.4}Ge_4$, $CeNi_8CuGe_4$, $Ce_{0.8}La_{0.2}Ni_9Ge_4$, and $LaNi_9Ge_4$ were synthesized by high frequency induction melting on a water cooled copper hearth under a protective argon atmosphere. The starting materials were Ce and La metals (Ames MPC [15], 99.95%), Ni (Johnson-Matthey, GB, 99.999%) and zone-refined Ge ingots (Johnson-Matthey, GB, 99.9999%). In a first step, Ni and Ge were melted together to produce a master alloy, which was then alloyed with Ce metal. The samples were finally annealed for one week at 1270 K in evacuated quartz ampules. Standard x-ray diffraction performed on powder revealed essentially phase pure samples crystallizing in the tetragonal space group $I4/mcm$ where substitution of Ni by Cu leads to a linear volume increase reaching 0.8% for $CeNi_8CuGe_4$ as compared to $CeNi_9Ge_4$. Details of the structural characterization were reported earlier [13].

Cold neutron quasi-elastic scattering experiments were carried out on the IN6 time-of-flight spectrometer at ILL Grenoble. The spectrometer was operated with an incident neutron energy of 3.15 meV (wavelength λ = 5.12 Å), yielding an energy resolution of 70 μeV at full width half maximum (FWHM). Powder samples, each with masses near 20 g, were enclosed in flat Al-can sample holders yielding an effective sample thickness of about 1.3 mm which was taken into account for proper absorption corrections. For calibration we measured a vanadium reference sample of the same disc-shape geometry at 2 K. For subtracting the background signal of the sample holder and intrinsic non-magnetic contributions to the total scattering (mainly phononic ones), we measured $LaNi_9Ge_4$ reference sample at same experimental conditions and

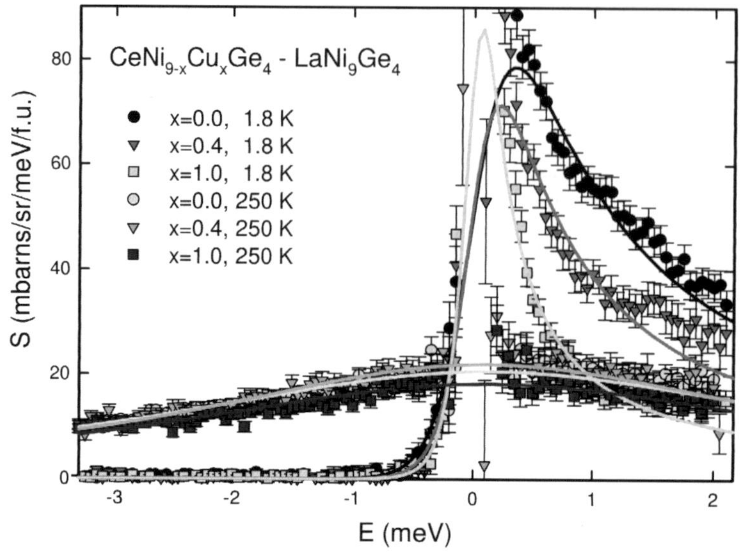

Figure 2. The magnetic correlation function $S_m(\omega)$ at 1.8 K and 250 K for compositions $CeNi_{9-x}Cu_xGe_4$ as labeled; solid lines are Lorentzian fits according to equation 2.

temperatures.

The muon spin relaxation (μSR) experiments at temperatures down to 35 mK were performed on the *MuSR* spectrometer at the ISIS facility where pulses of muons are implanted into the sample at 50 Hz and with a FWHM of 70 ns. These muons are thermalized in the bulk of the sample within few ps and decay with a half-life, $\tau_\mu = 2.2\,\mu$s into positrons, which are emitted preferentially in the direction of the muon spin axis. Each positron is time stamped and therefore the muon spin polarisation which corresponds to the asymmetry between positron counts in the forward and backward detectors (in so-called longitudinal geometry) can be followed as a function of time. In the present experiments, an initial asymmetry of 0.28 refers to an initially perfect spin polarization of the implanted muons parallel to the beam forward direction.

3. Cold neutron quasi-elastic studies

The variation of the Kondo temperature in the solid solution $CeNi_{9-x}Cu_xGe_4$ is evaluated from the quasi-elastic paramagnetic response probed by cold neutron scattering, i.e. via measurements of correlation functions $S(q,\omega)$ as described earlier in reference [8]. As the coherent and incoherent cross sections of $CeNi_{9-x}Cu_xGe_4$ with $x \leq 1$ and $LaNi_9Ge_4$ match each other within a few percent, the magnetic scattering $S_m(q,\omega)$ of cerium is obtained by subtracting the $LaNi_9Ge_4$ data representing phonon plus the background contributions. The magnetic scattering data $S_m(q,\omega,T \geq 1.8\text{K})$ obtained thereby exhibit only a weak q-dependence of the intensity which corresponds well to the usual Ce^{3+} form factor. Accordingly, q-integrated (section from 0.4 to 1.0 Å^{-1}) magnetic correlation functions $S_m(\omega)$ of $CeNi_9Ge_4$, $CeNi_{8.6}Cu_{0.4}Ge_4$, and $CeNi_8CuGe_4$ measured at 1.8 K and 250 K are displayed in figure 2. The magnetic scattering,

$$S_m(\omega,T) = \frac{\chi''(\omega,T)}{1-\exp(-\hbar\omega/k_BT)}, \tag{1}$$

is related to the absorptive component $\chi''(\omega,T)$ of the dynamic susceptibility which in case of a normal paramagnetic response with a single exponential decay of the magnetisation density is given by a Lorentzian

$$\chi''(\omega,T) = \chi'(T)\frac{\omega\Gamma(T)}{\omega^2+\Gamma^2(T)} \tag{2}$$

where $\Gamma(T)$ and $\chi'(T)$ are the temperature dependent relaxation rate (line width) and static susceptibility, respectively. Despite of an underlying CF splitting, the quasi-elastic response is acceptably well described by simple Lorentzian fits according to equation 2. While all the high temperature data measured at 250 K where k_BT is larger than the overall CF splitting are essentially on top of each other, there are rather significant changes of the low temperature quasi-elastic response.

The Lorentzian fits indicated by solid lines in figure 2 reveal a reduction of the line width Γ(1.8K) from 0.47 meV to 0.36 meV and 0.20 meV for $CeNi_9Ge_4$, $CeNi_{8.6}Cu_{0.4}Ge_4$, and $CeNi_8CuGe_4$, respectively. Simultaneously, also the overall integrated intensity which corresponds to the static susceptibility χ'(1.8K) (see equation 2) decreases from 0.104 emu/mol to 0.088 emu/mol, and 0.073 emu/mol, respectively. Although the trend is similar the significant variation of χ'(1.8K) seems in contradiction with the SQUID susceptibility data in figure 1b showing a minor variation of χ(1.8K). A close agreement of the value proposed by the Lorentzian fit with the SQUID result is observed for $CeNi_9Ge_4$ (from CF parameters in reference [8] 96% of the total magnetic response is expected from the $\Gamma_7^{(1)}$-$\Gamma_7^{(2)}$ quasi-quartet), but an increasing lack of quasi-elastic intensity as compared to the static SQUID susceptibility in the case of $CeNi_{8.6}Cu_{0.4}Ge_4$, and $CeNi_8CuGe_4$. The latter is obviously connected with a growing splitting of the $\Gamma_7^{(1)}$ and $\Gamma_7^{(2)}$ doublets caused by Cu substitution which alters the local environment of the cerium ions and, thereby, changes their CF scheme. The quasi-elastic neutron data, thus, reveal that Ni/Cu substitution causes a significant reduction of the Kondo energy scale and also a reduction of the effective ground state degeneracy towards a doublet in the case of $CeNi_8CuGe_4$. The latter is also confirmed by inelastic neutron data measured recently with an incident neutron energy of E_i = 20.5 meV at FRMII where an approximate CF scheme with doublets at ground state, 5 meV and 12 meV is observed for $CeNi_8CuGe_4$ [16].

4. Muon spin relaxation studies

In order to probe magnetic correlations at low (dilution fridge) temperatures we utilized the μSR technique which is a local probe method sensitive to extremely small internal fields and ideally suited to detect any kind of static magnetism as well as dynamic magnetic features (if accessible in μs time scale of the muon life time). The μSR technique has, thus, been extensively applied to heavy fermion systems (see e.g. the review by Amato [17]).

The zero-field μSR measurements down to about 40 mK for $CeNi_9Ge_4$, $CeNi_{8.6}Cu_{0.4}Ge_4$, $CeNi_8CuGe_4$, and a magnetically dilute sample $Ce_{0.8}La_{0.2}Ni_9Ge_4$ are displayed in figure 3 as time dependent depolarization of the muon spins, i.e. asymmetry estimated from the positions detected in beam forward and beam backward directions. All relaxation data measured at highest temperatures displayed in figure 3, i.e. at 4 – 5 K, are in approximate agreement with the Gaussian Kubo-Toyabe relaxation function (see e.g. reference [17]) which refers a depolarisation of the muon spins by quasi-static nuclear dipolar fields. As the largest nuclear moments are contributed by copper atoms, $CeNi_8CuGe_4$ in figure 3a clearly displays the quickest relaxation at high temperatures and nuclear dipolar fields maintain their dominant role down to 0.3 K. Just near 200 mK, where the susceptibility in figure 1b exhibits a sharp cusp indicating the onset of AF order, there is a dramatic change of the μSR signal showing a rather quick depolarization towards a typical background asymmetry $A_{bg} \sim 0.05$ at low temperatures. The absence of any sign of coherent frequency oscillations in the zero field μSR spectra which has been observed also in some other antiferromagnetic materials, e.g. $Ce_8Pd_{24}Ga$ [19], refers to a muon stopping site in a high symmetry position with respect to the Ce sublattice where dipolar fields of AF

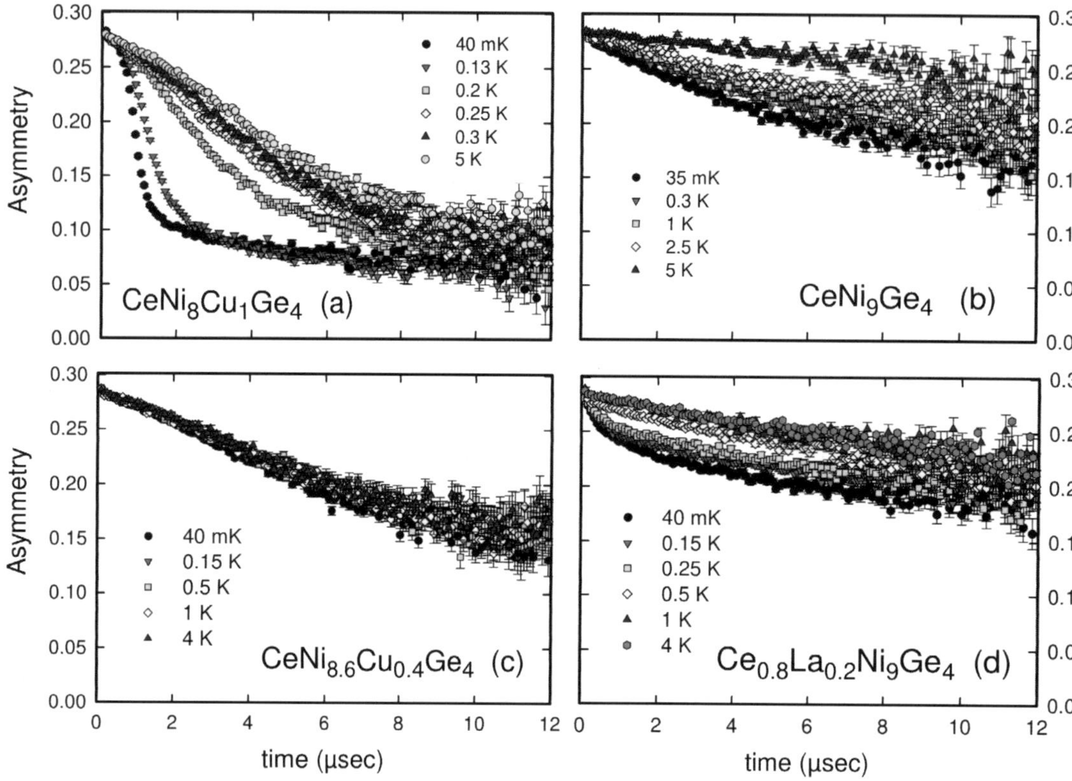

Figure 3. The time dependent μSR asymmetry data of $CeNi_8CuGe_4$ (a), $CeNi_9Ge_4$ (b), $CeNi_{8.6}Cu_{0.4}Ge_4$ (c), and $Ce_{0.8}La_{0.2}Ni_9Ge_4$ (d) at selected temperatures and zero field.

ordered Ce moments may compensate each other to zero. In the presence of substitutional disorder in $CeNi_8CuGe_4$, muon spins at such point of compensating dipolar fields should, of course, experience some distribution of static internal fields with a zero mean at the muon site. This interpretation of the $CeNi_8CuGe_4$ 40 mK data is supported by a markedly different field dependence of $CeNi_8CuGe_4$ observed in longitudinal field μSR measurements (not shown for the sake of brevity) as compared to all other compounds studied in this investigation.

The exponential low temperature relaxation of $CeNi_9Ge_4$ and magnetically dilute $Ce_{0.8}La_{0.2}Ni_9Ge_4$ in figure 3b and 3d refers a muon spin depolarization by dynamically fluctuating dipolar fields. In the case of $CeNi_9Ge_4$ a clean simple exponential behaviour $A(t) \propto \exp(-\lambda t)$ with a depolarization rate $\lambda = 0.133\,\mu s^{-1}$ at 35 mK is observed (see figure 3b). As already discussed in reference [8] a simple exponential μSR signal rules out a disorder or inhomogeneity dominated NFL regime which would be indicated by a broad distribution and a strong temperature dependence of μSR relaxation rates [18]. The latter situation seems to apply for the solid solution $Ce_{0.8}La_{0.2}Ni_9Ge_4$ which exhibits a significantly larger initial relaxation rate and strongly stretched exponential behaviour. The larger initial relaxation rate is counter-intuitive for the magnetically dilute case in particular when keeping in mind the approximate concentration scaling of magnetic contributions to the specific heat and susceptibility reported by Killer *et al.* [7]. These effects of magnetic dilution are even more remarkable when considering the corresponding results of $CeNi_{8.6}Cu_{0.4}Ge_4$ which according to our thermodynamic studies [13] is located right at the critical concentration for the onset of AF order and clearly exhibits quantum critical behavior with χ and $C/T \propto -\ln T$. Against all expectations μSR results in figure 3c do not reveal any temperature dependence of the $CeNi_{8.6}Cu_{0.4}Ge_4$ data even from 40 mK to 4 K where muon spins essentially probe nothing else than the nuclear dipolar fields of the copper cores.

The only plausible interpretation of these puzzling results are strong AF short range magnetic correlations causing a compensation dipolar fields at the muon site which is most effective right at the quantum critical point, i.e. muon will not see any change with temperature because nuclear dipolar fields remain dominant at all temperatures. Magnetic dilution through Ce/La substitution of course creates at least two kinds of muon stopping sites: (i) in between a pair of Ce-ions where correlated $4f$ moments may compensate each other and (ii) in between a pair of Ce and La where the muon spin feels the full Ce moment. Longitudinal field μSR data (not shown) in fact support this scenario where a smaller number of muon sites ($\leq 40\%$) is of the second type and causes the quick initial relaxation and is more robust against the externally applied longitudinal field, i.e., connected with larger local fields at the muon site, and a larger number of muon sites is of type (i) displaying a much slower relaxation which is easily suppressed by just a few 10 G longitudinal field similar to the field dependence seen in $CeNi_9Ge_4$ and $CeNi_{8.6}Cu_{0.4}Ge_4$.

5. Summary

The effectively quasi-fourfold ground state degeneracy in combination with a Kondo temperature of only a few Kelvin places the paramagnetic heavy fermion system $CeNi_9Ge_4$ in an exceptional position among Kondo lattice systems where the paramagnetic ground state could be stabilized by the quasi-quartet CF ground state. The latter is suggested by the fact that tuning $CeNi_9Ge_4$ towards a magnetic ground state is accomplished by Ni/Cu as well as Ni/Co substitution (equivalent to electron and hole doping, respectively) which are both leading to a slightly modified local environment of the cerium ions and, thereby, cause a reduction of the CF ground state towards a doublet one. These modifications of ground state degrees of freedom are revealed by microscopic studies via cold neutron quasi-elastic scattering on $CeNi_9Ge_4$ and the solid solutions $CeNi_{8.6}Cu_{0.4}Ge_4$ and $CeNi_8CuGe_4$ where a significant narrowing of the quasi-elastic line at low temperatures and simultaneously some loss in quasi-elastic intensity is observed. The latter indicates a transfer of quasi-elastic to inelastic scattering caused by a splitting of the quasi-quartet CF ground state towards two well separated doublets.

Local probe μSR studies performed down to temperatures near 40 mK revealed rather unexpected results where on the one hand, magnetic dilution (Ce/La substitution) tends to increase the local fields experienced by the implanted muons, whereas on the other hand, tuning towards quantum criticality in $CeNi_{8.6}Cu_{0.4}Ge_4$ tends to annihilate the dipolar fields of cerium moments at the muon stopping sites. A cancellation of the dipolar fields of cerium $4f$ moments is only possible if muons locate in a high symmetry position with respect to the Ce sublattice where strong AF correlations lead to a pairwise compensation of cerium dipolar fields. The μSR results are, thus, suggestive of strong AF short range dynamic correlations being present in the unusual Kondo lattice ground state of $CeNi_9Ge_4$.

Acknowledgments

This research project was supported by COST P-16 ECOM. The μSR studies at the ISIS facility were supported by the European Commission under the 7th Framework Programme through the Key Action: Strengthening the European Research Area, Research Infrastructures (Contract CP-CSA-INFRA-2008-1.1.1 Number 226507-NMI3).

The authors thank the Yukawa Institute for Theoretical Physics at Kyoto University. Discussions during the YITP workshop YITP-W-10-12 on "International and Interdisciplinary Workshop on Novel Phenomena in Integrated Complex Sciences: from Non-living to Living Systems" were useful to complete this work.

References

[1] Mathur M D, Grosche F M, Julian S R, Walker I R, Freye D M, Hasselwimmer R and Lonzarich G G 1998 *Nature* **394** 39
[2] Saxena S S, Agarwal P, Ahilan K, Grosche F M, Hasselwimmer R, Steiner M J, Pugh E, Walker I R, Julian S R, Monthough P, Lonzarich G G, Huxley A, Sheikin I, Braithwaite D and Flouquet J 2000 *Nature* **406** 587
[3] Sachdev S 2000 *Quantum Phase Transitions* (Cambridge University Press)
[4] Stewart G R 2001 *Rev. Mod. Phys.* **73** 797; Stewart G R 2006 *Rev. Mod. Phys.* **78** 743
[5] von Löhneysen H, Rosch A, Vojta M and Wölfle P 2007 *Rev. Mod. Phys.* **79** 1015
[6] Michor H, Bauer E, Dusek C, Hilscher G, Rogl P, Chevalier B, Etourneau J, Giester G, Killer U and Scheidt E–W 2004 *J. Magn. Magn. Mater.* **272-276** 227
[7] Killer U, Scheidt E–W, Eickerling G, Michor H, Sereni J, Pruschke T and Kehrein S 2004 *Phys. Rev. Lett.* **93** 216404
[8] Michor H, Adroja D T, Bauer E, Bewley R, Dobozanov D, Hillier A D, Hilscher G, Killer U, Koza M, Manalo S, Manuel P, Reissner M, Rogl P, Rotter M and Scheidt E–W 2006 *Physica* B **378-380** 640
[9] Anders F and Pruschke T 2006 *Phys. Rev. Lett.* **96** 086404
[10] Scheidt E–W, Mayr F, Killer U, Scherer W, Michor H, Bauer E, Kehrein S, Pruschke T and Anders F 2006 *Physica* B **378-380** 154
[11] Sereni J 1991 *Handbook on the Physics and Chemistry of Rare Earths* vol 15 ed K A Gschneidner Jr and L Eyring (Amsterdam: North-Holland) pp 1-59
[12] Coleman P 1983 *Phys. Rev.* B **28** 5255
[13] Peyker L, Gold C, Scheidt E–W, Scherer W, Donath J G, Gegenwart P, Mayr F, Unruh T, Eyert V, Bauer E and Michor H 2009 *J. Phys.: Condens. Matter* **21** 235604
[14] Peyker L, Gold C, Scherer W, Michor H and Scheidt E-W 2011 *J. Phys.: Conf. Series* **273** 012049
[15] Materials Preparation Center, Ames Laboratory, US DOE Basic Energy Sciences, Ames, IA, USA, available from *www.mpc.ameslab.gov*
[16] Peyker L, Gold C, Scherer W, Michor H, Unruh T, Simeoni G G, Senyshyn A, Adroja D T, Stockert O and Scheidt E-W 2011 *EPL* **93** 37006
[17] Amato A 1997 *Rev. Mod. Phys.* **69** 1119
[18] MacLaughlin D E, Heffner R H, Bernal O O, Ishida K, Sonier J E, Nieuwenhuys G J, Maple M B and G R Stewart 2004 *J. Phys.: Condens. Matter* **16** S4479
[19] Adroja D T, Kockelmann W, Hillier A D, So J Y, Knight K S and Rainford B D 2003 *Phys. Rev.* B **67** 134419

Magnetic Isotherm of Itinerant Electron Magnets – A New Approach to Itinerant Electron Metamagnetism

Yoshinori Takahashi

Graduate School of Material Science, University of Hyogo, Kouto 3-2-1, Kamigori, Ako, Hyogo 678-1297, JAPAN

E-mail: `takahash@sci.u-hyogo.ac.jp`

Abstract. Theoretical development on the magnetic isotherm in itinerant electron ferromagnets is briefly reviewed in comparison with observed magnetization measurements. Itinerant electron metamagnetic transitions originate from the magnetization curve derived from the negative fourth expansion coefficient of the free energy in magnetization. In our view, however, it is difficult to expect the negative coefficient of the magnetic origin. We therefor propose a new mechanism of the metamagnetic transition, by taking an explicit account of volume dependence of the free energy, which is consistent with our spin fluctuation theory of magnetic isotherms.

1. Introduction

Metamagnetism is usually observed in highly exchange enhanced itinerant electron paramagnets close to the magnetic instability point. It exhibits as a sudden discontinuous jump of the induced magnetic moment at some critical magnetic field H_c. According to the widely accepted theoretical idea, the phenomenon originates from the distinct magnetic isotherm associated with the concave shape of density of states of single-electron excitations around the Fermi energy. At finite temperatures, transitions are affected by thermal spin fluctuations. On the other hand, magnetic isotherms are exclusively determined by the effect of spin fluctuations influenced by the externally applied magnetic field. This can be applied even in the ground state with no thermal spin fluctuation amplitudes. This means that the theoretical bases of metamagnetic transitions conflicts with the spin fluctuation theory of magnetic isotherms.

The motivation of this study is, therefore, first to make a brief review of the current theoretical ideas on the magnetic isotherm of itinerant electron magnets. We also show how the theoretical predictions are verified experimentally. We next propose a new idea for the metamagnetic transition observed in itinerant electron magnets, that is consistent with the spin fluctuation theory.

2. Theories of magnetic isotherm of itinerant electron magnets

If we apply magnetic field H to magnets, the magnetic moment M is induced. The relation between H and M, called magnetic isotherm, is usually described by the following thermodynamic relation:

$$H = \frac{\partial F(M,T)}{\partial M} = a(T)M + b(T)M^3 + \cdots . \quad (1)$$

This is a fundamental property of magnetic materials, for we can derive a number of magnetic properties from them. One of the main concerns of theoretical studies has been therefore to determine the magnetic isotherm as the effect of spin fluctuations in the presence of the magnetic field. It has long been believed that observed magnetization curves in itinerant ferromagnets are well-fitted with the above formula with almost temperature independent $b(T)$ as far as the induced static moment is very small. Only the temperature dependence of the coefficient $a(T)$ of the first term in the right hand side was a subject of our main interest. However, the idea is now regarded as too simple and more elaborate treatments are necessary.

Theory of magnetic isotherms is crucial for our understanding of the mechanism of metamagnetic transitions. At present there are following two opposite theoretical explanations for them.

(i) Band splitting mechanism
According to this picture, the splitting of energies of each conduction electron states induced by the applied magnetic field is the origin of the induced magnetic moment in the crystal. The well-known self-consistent renormalization (SCR) theory of spin fluctuations is also classified as this, because it is equivalent with the single-particle band picture as far as the isotherm in the ground state is concerned.

(ii) Spin fluctuation mechanism
In this picture, amplitudes of spin fluctuations suppressed by the external magnetic field is regarded as the origin of induced magnetic moment.

Depending on mechanisms, we expect a different temperature dependence of $b(T)$. The first view gives the very weak dependence, that can be usually neglected. On the other hand, $b(T)$ becomes temperature dependent in the second view. Most of experiments to date support the second view.

2.1. Spin fluctuation theory of itinerant electron magnets

What follow are basic mechanisms how magnetic properties are determined by the effects of spin fluctuations. They are based on the following basic two ideas [1, 2].

(i) The total spin amplitude conservation (TAC)
This means that the sum of the squared quantum and thermal spin fluctuation amplitudes and the squared static moment remains constant independent of temperature and/or in the presence of the external magnetic field.

(ii) The global consistency (GC) in the H-M space
The SCR spin fluctuation theory is mainly concerned with the temperature dependence of $a(T)$ of the first term of the isotherm (1), i.e., the inverse of magnetic susceptibility. We mean by GC, on the other hand, that we should rather deal with the relation between H and M over a wide range of their variables.

For paramagnets, for simplicity, the equal-time local spin correlation is given by

$$\langle \mathbf{S}_{loc}^2 \rangle = \frac{3}{N_0^2} \sum_q \int_0^\infty \frac{d\omega}{\pi} \coth(\omega/2k_BT)\mathrm{Im}\chi(\mathbf{q},\omega), \tag{2}$$

where the imaginary part of the dynamical magnetic susceptibility is given in the following double Lorentzian form,

$$\begin{aligned} \mathrm{Im}\chi(\mathbf{q},\omega) &= \chi(\mathbf{q},0)\frac{\omega\Gamma_q}{\omega^2+\Gamma_q^2}, \\ \chi(\mathbf{q},0) &= \chi(0,0)\frac{\kappa^2}{q^2+\kappa^2}, \quad \Gamma_q = \Gamma_0 q(\kappa^2+q^2). \end{aligned} \tag{3}$$

The number of magnetic atoms, the correlation wave-number squared, and the damping constant are, respectively, denoted by N_0, κ^2, and Γ_q. From the decomposition, $\coth(\omega/2k_BT) = 1 + 2n(\omega)$, the amplitude in (2) is split into thermal and zero-point amplitudes as follows.

$$\begin{aligned} 3A_t(\kappa^2, T) \equiv \langle \mathbf{S}_{loc}^2 \rangle_\mathrm{T} &= \frac{6}{N_0^2} \sum_q \int_0^\infty \frac{d\omega}{\pi} n(\omega) \mathrm{Im}\chi(\mathbf{q}, \omega), \\ 3A_z(\kappa^2) \equiv \langle \mathbf{S}_{loc}^2 \rangle_\mathrm{Z} &= \frac{3}{N_0^2} \sum_q \int_0^\infty \frac{d\omega}{\pi} \mathrm{Im}\chi(\mathbf{q}, \omega). \end{aligned} \tag{4}$$

Both amplitudes depend on κ^2 through its dependence of $\chi(q, 0)$ and Γ_q in (3). The thermal amplitude has an extra explicit temperature dependence through the Bose distribution function $n(\omega)$. In the presence of the spontaneous or induced magnetic moment due to external magnetic field, κ^2 becomes anisotropic, as given by $\kappa_\parallel^2 \propto \partial H/\partial M$ and $\kappa_\perp^2 \propto H/M$ for fluctuations parallel or perpendicular to the direction of the static moment. The amplitude conservation can now be written in the form,

$$[2A_t(\kappa_\perp^2, T) + A_t(\kappa_\parallel^2, T)] + [2A_z(\kappa_\perp^2) + A_z(\kappa_\parallel^2)] + m^2 = 3[A_t(0, T_c) + A_z(0)], \tag{5}$$

where $m = M/N_0$ is the magnetic moment per magnetic atom. The right hand side represents the value of the amplitude at the critical point.

We require that (5) should be regarded as a differential equation of H as a function of M, for we can find the derivative, $\partial H/\partial M$, for given values of M and H from (5). Because M dependence of H is obtained as a solution of (5), our main concerns are not restricted to the first term of (1) but the global behavior of M dependence over the wide range of the external magnetic field. This is what we mean by GC.

For example, the thermal amplitude increases with temperature in the paramagnetic phase. To satisfy the amplitude conservation,

$$A_t(\kappa^2, T) + A_z(\kappa^2) - A_z(0) = A(0, T_c). \tag{6}$$

the zero-point amplitude has to be suppressed to compensate for the increase of the thermal amplitude by the temperature dependent magnetic susceptibility, $\chi \propto \kappa^{-2}$. As a solution of (6), the temperature dependence of κ^2, i.e. the Curie-Weiss law behavior of the magnetic susceptibility, is derived. In the same way, the spin amplitude suppression by the external magnetic field induces a static magnetic moment to satisfy the TAC requirement, that determines the magnetic isotherm. The mechanism is quite different from that of the band picture.

2.2. Theory of Magnetic isotherm

According to the spin fluctuation mechanism, various interesting magnetic properties have been derived. We are particularly concerned with only three of them, which are all related to the magnetic isotherm.

(i) Magnetic isotherm in the ground states
The first example is related to the magnetization curve in the ground state. It can be determined by solving the differential equation,

$$2[A_z(\kappa_\perp^2) - A_z(0)] + [A_z(\kappa_\parallel^2) - A_z(0)] + m^2 = A_t(0, T_c). \tag{7}$$

Since $A_z(\kappa^2) - A_z(0) \propto \kappa^2$ for $\kappa^2 \ll 1$ and from the relations $\kappa_\perp^2 \propto H/M$, $\kappa_\parallel^2 \propto \partial H/\partial M$, we can easily find the solution in the form,

$$H = a_0 M + b_0 M^3 \tag{8}$$

The substitution into (7) gives the following b_0 of the second term in the right hand side given by

$$b_0 = \frac{4k_B T_A^2}{15(2\mu_B)^4 N_0^3 T_0}, \tag{9}$$

where T_0 and T_A, defined by $\Gamma_0 q_B^3/2\pi$ and $q_B^2/[2\chi(0,0)\kappa^2]$ with zone boundary wave-vector q_B, represent the dispersions of spin fluctuation spectra in frequency and wave-vector spaces.

(ii) Critical magnetic isotherms
In the critical limit, since $\kappa^2 \to 0$ as $T \to T_c$, (5) is written in the form,

$$2[A_t(\kappa_\perp^2, T_c) - A_t(0, T_c)] + [A_t(\kappa_\parallel^2, T_c) - A_t(0, T_c)] + m^2 = 0, \tag{10}$$

because the effects of zero-point fluctuations become negligible in this limit. From the critical behavior of the thermal amplitude, $A_t(\kappa^2, T) - A_t(0, T) \propto \sqrt{\kappa^2}$, we can find the solution of the critical magnetic isotherm of the form,

$$H \propto M^5. \tag{11}$$

The moment M is induced in proportional to $H^{1/5}$ at $T = T_c$, instead of $H^{1/3}$ by the band picture.

(iii) Positive mode-mode coupling in semiconducting paramagnets
The coefficient $b(T)$ in (1) is regarded as the nonlinear mode-mode coupling among spin fluctuation modes. While the negative sign is derived by the band splitting mechanism, we can predict the positive value even in cases of semiconducting paramagnets with concave form of the density of states around the Fermi energy [3].

We now show how these properties are verified experimentally in the following subsections.

2.3. Magnetic isotherm in the ground state

In place of the Rhodes-Wohlfarth plot (p_C/p_s vs T_c), a new revised plot, $p_{\rm eff}/p_s$ vs T_c/T_0, was proposed by Takahashi [1, 2], where p_s, $p_{\rm eff}$ and p_C represent the spontaneous moment in the low temperature limit, the effective moment estimated from the Curie constant of the magnetic susceptibility, and $p_C(p_C + 2) = p_{\rm eff}^2$. It is based on the universal relation between these two ratios that can be derived from (9) in the ground state. As shown in Figure 1, almost all the experimentally estimated ratios fall near the solid curve determined theoretically. The good agreement of the plot is equivalent with the validity of (9). It is very difficult to figure out its reason based on the band picture.

2.4. Critical magnetic isotherm

The idea of critical magnetic isotherm had been long absent in the theories of itinerant magnetism. Since the relation (11) was first predicted by Takahashi [1], the observed linear relations satisfied between M^4 and H/M have now accumulated. As an old example, we show in Figure 2 (above) the observed values of M^4 almost at the critical point, $T = 29$ K, plotted against H/M. The conventional Arrott plot [4] is also shown in the same figure (below). Recently the good linear relations are observed by Nishihara et al. in Ni [5] and the Heusler compound Co_2CrGa [6]. These results clearly shows that coefficient b of the free energy must be temperature dependent and has to vanish at the critical point, in contradiction to the band picture and the assumption of the SCR theory.

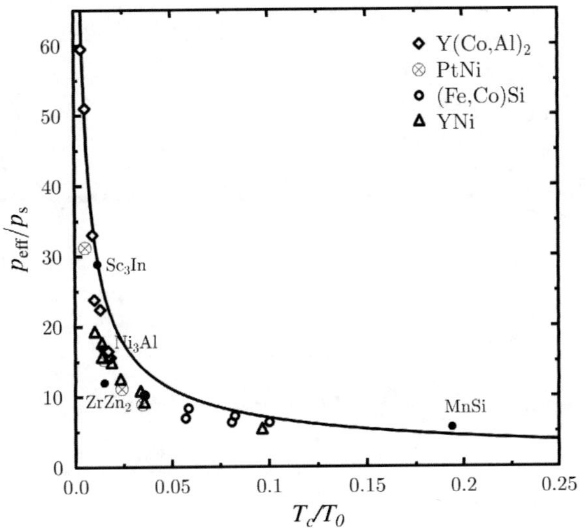

Figure 1. Takahashi-Rhodes-Wohlfarth plot for itinerant electron ferromagnets. Solid circles are for MnSi, Ni_3Al, $ZrZn_2$, and Sc_3In.

2.5. Sign of mode-mode coupling constant in semiconducting paramagnets

The last example is closely related to the metamagnetic transitions. According to the band picture, the coefficient $b(T)$ is given by

$$b = \frac{1}{\rho^3}\left[\left(\frac{\rho'}{\rho}\right)^2 - \frac{\rho''}{3\rho}\right]. \tag{12}$$

For semiconducting materials with the concave shape of the density of states curve ρ around the Fermi energy, b becomes negative, i.e., in cases of $\rho'' > 0$ and $\rho' \sim 0$. Analyses of the magnetization process of semiconducting compounds, FeSi by Koyama [7] and $FeSb_2$ by Koyama [8], show that the couplings b determined from their slopes in the Arrott plots are rather positive, that favors again the spin fluctuation mechanism even qualitatively.

3. Theory of metamagnetic transitions

Let us now turn to the subject of metamagnetic transitions. The current dominant idea of itinerant metamagnetic transition is based on the following magnetic isotherm [9],

$$H = a(T)M + b(T)M^3 + cM^5. \tag{13}$$

The coefficient $b(T)$ in the second term is assumed to be negative because of the concave form of the density of states around the Fermi energy. However as we have shown in the preceding section, the negative $b(T)$ of the purely magnetic origin is very difficult from the spin fluctuation mechanism.

In addition, the third M^5 term with finite positive c is necessary in (13) for the stability of the free energy. Only because of this higher order nonlinear mode-mode coupling, $b(T)$ becomes temperature dependent and changes its sign at high temperatures. There however

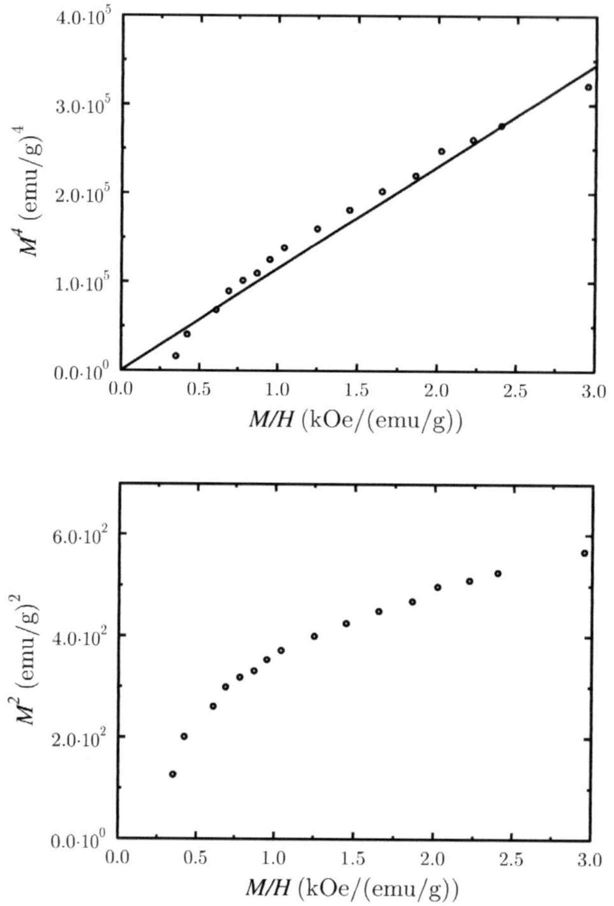

Figure 2. Magnetization curve of MnSi at $T = 29$ K reploted in the form, M^4-H/M, (upper) and in Arrott plot (lower).

exist unfavorable experiments for this term. We show the Arrott plot of high field magnetic measurements on $Y(Co_{0.91}Al_{0.09})_2$ by Goto and Bartashevich (1998) [11] in Figure 3. In their original figure, induced moment M is plotted against H. The linear relation seems to be satisfied between M^2 and H/M in the ferromagnetic state after the transition. The same behavior is also observed recently in the magnetization curve of $LaCo_9Si_4$ by Michor et al [12] in even wider range of H/M. All of these observations clearly require a new idea for more plausible understanding of the phenomena.

3.1. Volume dependence of the magnetic isotherm

As a new mechanism, it seems necessary to introduce another degrees of freedom that couples with magnetism. The volume will be one of the best candidates, for the increase or the appearance of the magnetization usually accompanies the volume expansion of crystals. The metamagnetic transitions will then occur between two almost degenerate free energies, $F_p(M,V)$ and $F_f(M,V)$, from the paramagnetic state, stable around the volume V_0, to the ferromagnetic

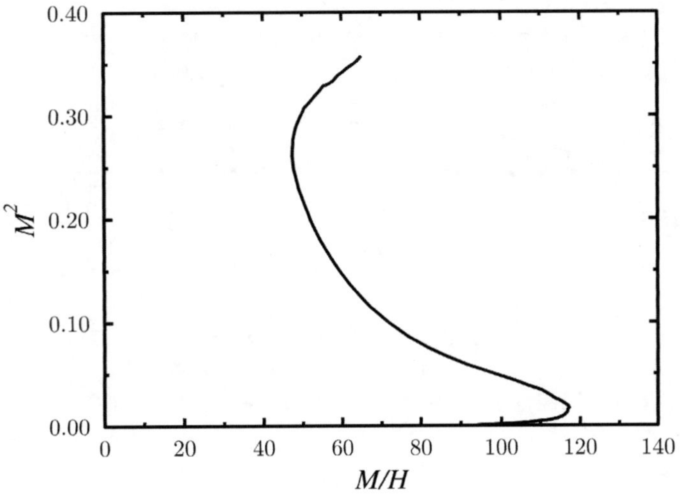

Figure 3. Arrott plot of the metamagnetic transition in $Y(Co_{0.91}Al_{0.09})_2$.

state around the volume V_1 in the presence of magnetic field. The magneto-volume effect [10] is assumed to be involved in the transition. In the paramagnetic ground state for $V \sim V_0$, the free energy is expanded in M as follows.

$$F_p(M,V) = F(0,V) + \frac{1}{2}a(V)M^2 + \frac{1}{4}b(V)M^4 \tag{14}$$

In the absence of H, both $a(V_0) > 0$ and $b(V_0) > 0$ are satisfied for volume $V = V_0$. For ferromagnetic state, metastable around $V \sim V_1$, it is given by

$$\begin{aligned} F_f(M,V) &= F_f(M_0,V) + \frac{1}{2}a(V)[M^2 - M_0^2] + \frac{1}{4}b(V)[M^4 - M_0^4] \\ &= F_f(M_0,V) - a(V)[M - M_0]^2 + b(V)M_0[M - M_0]^3 + \cdots, \\ F_f(M_0,V) &= F(0,V) + \frac{1}{2}a(V)M_0^2 + \frac{1}{4}b(V)M_0^4, \end{aligned} \tag{15}$$

where $M_0^2 = -a(V)/b(V)$ and $a(V_1) < 0$.

Because of the volume dependence of expansion coefficients around V_0, $a(V)$ is given by

$$a(V) \simeq a(V_0) + a'(V_0)\delta V = a(V_0) - \frac{1}{2}\gamma_p M^2,$$
$$\delta V = V - V_0 = V_0 C_p M^2, \quad \gamma_p = -2V_0 a'(V_0) C_p$$

where C_p is the forced magneto-volume coupling constant. Because of the forced magneto-volume effect, the volume change is equivalent with the induced magnetic moment squared. This means that $b(V)$ in (14) effectively becomes negative if

$$b(V) = b(V_0) - \gamma_p < 0,$$

is satisfied. The point is that the negative b becomes effectively possible because of the magneto-volume effect.

Journal of Physics: Conference Series **344** (2012) 012002 doi:10.1088/1742-6596/344/1/012002

3.2. Condition of the transition

Once b becomes negative, metamagnetic transitions in the ground state can be realized as follows. For field-increasing processes, the transition occurs at the field H_c of divergent differential magnetic susceptibility, $\partial M/\partial H = \infty$, or zero of its reciprocal, $\partial H/\partial M = a + bM^3 = 0$. It follows then,

$$M_{c1}^2 = \frac{a}{3|b|}, \quad H_c = -2bM_{c1}^3 = \frac{2a}{3^{3/2}}\sqrt{\frac{a}{|b|}} = \frac{2}{3}aM_{c1}.$$

The transition occurs from the state with the moment M_{c1} and the volume V_0 slightly shifted by $\delta V = C_p M_{c1}^2$. The ratio of the critical field H_c and the moment M_{c1} is given by

$$\frac{H_c}{M_{c1}} = \frac{2}{3}a = \frac{2}{3}\frac{H}{M}\bigg|_{H\to 0}. \tag{16}$$

The reverse process with decreasing field is also realized in the same way. If the transition occurs from the moment M_{c1} to M_{c2} at the critical field $H = H_c$, the magnetic isotherm of the ferromagnetic state can be expanded as follows.

$$H = H_c + \alpha(M - M_{c2}) + \beta(M - M_{c2})^2 + \cdots, \tag{17}$$

where α and β are coefficients renormalized by the effect of magneto-volume effects. The volume V_1 and M_{c2} are evaluated by solving the simultaneous equations,

$$\begin{aligned} H_c &= a(V_1)M_{c2} + b(V_1)M_{c2}^3, \\ \frac{\partial F(M_{c2}, V_1)}{\partial V_1} &= \frac{\partial F(0, V_1)}{\partial V_1} + H_c\frac{\partial M_{c2}}{\partial V_1} + \frac{1}{2}a'(V_1)M^2 + \frac{1}{4}b'(V_1)M_{c2}^4 = 0. \end{aligned} \tag{18}$$

The magnetic isotherm of the system with the volume around V_1 is then given by

$$\begin{aligned} H &= H_c + [a(V_1) + 3M_{c2}^2 b(V_1)]\delta M + 3M_{c2}b(V_1)\delta M^2 \\ &\quad + (a'M_{c2} + b'M_{c2}^3)\delta V + (a' + 3M_{c2}^2 b')\delta V\delta M + \cdots \\ \frac{\partial}{\partial V}&\left[F(0,V) + \frac{1}{2}a(V)M_{c2}^2 + \frac{1}{4}b(V)M_{c2}^4\right]\delta V \\ &= -H\frac{\partial M}{\partial V} + H_c\frac{\partial M_{c2}}{\partial V} - \frac{1}{2}a'(V)[M^2 - M_{c2}^2] - \frac{1}{4}b'(V)[M^4 - M_{c2}^4], \end{aligned} \tag{19}$$

where $\delta M = M - M_{c2}$. From the second equation of (19), the forced magnetostriction δV, and therefore the volume dependence of $a(V)$ are given by

$$\begin{aligned} \delta V &= V_1 C_f(M^2 - M_{c2}^2) \simeq 2V_1 C_f M_{c2}\delta M, \\ a(V) &\simeq a(V_1) + a'(V_1)\delta V = a(V_1) - \gamma_f M_{c2}\delta M, \end{aligned} \tag{20}$$

where the first two terms and the last term are neglected. The magneto-volume coupling in the ferromagnetic state is denoted by C_f and $\gamma_f = -2V_1 a'(V_1)C_f$. The substituting (20) into the first line of (19) leads to values of α and β in (17) given by

$$\alpha = a(V_1) + 3M_{c2}^2 b(V_1) - \gamma_f M_{c2}^2, \quad \beta = M_{c2}[3b(V_1) + \gamma_f]. \tag{21}$$

The critical field for the reverse transition is now determined by

$$\frac{\partial H}{\partial M} = \alpha + 2\beta(M - M_{c2}) = 0. \tag{22}$$

The critical field and the moment are given by

$$M_{c2}' = M_{c2} - \frac{\alpha}{2\beta}, \quad H_c' = H_c - \frac{\alpha^2}{4\beta} \tag{23}$$

Because of the magneto-volume effect, smaller α and larger β lead to the smaller difference between the critical fields, H_c and H_c'.

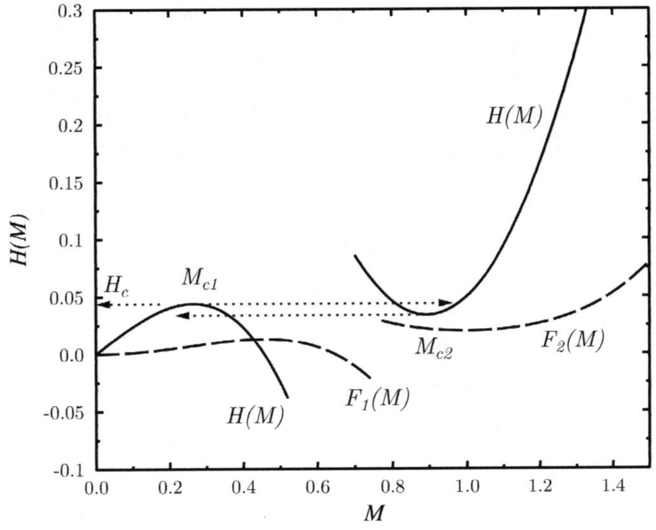

Figure 4. Free energies and magnetic isotherms, H vs M, of metamagnetic transitions.

The proposal of my idea of metamagnetic transition is clearly shown in Figure 4. In the left part, the magnetization curve and the paramagnetic free energy around the volume V_0 are shown. Those around V_1 shown in the right hand side are for the ferromagnetic state. These are effective magnetic isotherms and free energies. Transitions occur at the maximum and the minimum of H for increasing magnetic field. In our mechanism, we only need the conditions for the transition:

- b becomes effectively negative due to the volume effect, for instance.
- Systems must be very close to the instability point for transitions in lower magnetic field range.

The above idea is actually supported by the high field experiment of the metamagnetic transition on $Y(Co,Al)_2$ [11]. Simultaneously observed discontinuous volume changes at the transitions clearly show that both the volume change and the metamagnetic transition are intimately correlated with each other.

4. Summary

In this study we made a brief review on predominant rolls of spin fluctuations on the magnetic isotherm of itinerant electron magnets by paying particular attention on the following properties:

- Magnetization curves in the ground state
- Critical magnetic isotherms
- Positive fourth expansion coefficient of the free energy in magnetization observed in semiconducting paramagnets

Agreements between theoretical predictions and experiments demonstrate that magnetization curves of itinerant electron magnets are exclusively determined by the effect of spin fluctuations.

The above view imposes a stringent restriction to the magnetization curves. It becomes very difficult to assume a negative coefficient $b(T)$ of the isotherm in (1). We therefore propose a new idea as a basic mechanism of itinerant metamagnetic transition. It consists of the assumption that the negative b becomes possible only because of the coupling between magnetism and other degrees of freedom, a change of the crystal volume, for instance. The absence of the higher M^6 term in the free energy expansion observed in the magnetization curves of $Y(Co,Al)_2$ and $LaCo_9Si_4$ is consistent with our picture.

We are particularly concerned with the metamagnetic transition in the ground state by taking the forced magneto-volume effects into account. At finite temperature we also need to take account of the effect of thermal volume expansions. The temperature dependence of the magnetization curve and magnetic susceptibility will be the subject of our future studies including the detailed comparison with experiments as well.

Acknowledgments

This workshop was supported in part by the Grant-in-Aid for the Global COE Program "The Next Generation of Physics, Spun from Universality and Emergence" from the Ministry of Education, Culture, Sports, Science and Technology (MEXT) of Japan. The author thanks Professor H. Michor for valuable discussion on experiments of the metamagnetic transition and the Yukawa Institute for Theoretical Physics at Kyoto University, where this work was initiated during the YITP-W-10-12 on "International and Interdisciplinary Workshop on Novel Phenomena in Integrated Complex Sciences: from Non-living to Living Systems". This work was supported by a Grant-in-Aid for Science Research (C) 21540341.

References

[1] Takahashi Y 1986 *J. Phys. Soc. Japan* **55** 3553
[2] Takahashi Y 2001 *J. Phys.: Condens. Matter* **13** 6323
[3] Takahashi Y 1998 *J. Phys.: Condens. Matter* **10** L671
[4] Bloch D, Voiron J, Jaccarino, V and Wernick J H 1975 *Phys. Lett.* A **51** 259
[5] Nishihara H, Komiyama K, Oguro I, Kanomata T and Chernenko V 2007 *J. Alloys Compd.* **442** 191
[6] Nishihara H, Furutani Y, Wada T, Kanomata T, Kobayashi K, Kainuma R, Ishida K and Yamauchi T 2010 J. Phys: Conf. Ser. **200** 032053
[7] Koyama K, and Goto T, Kanomata T, Note R and Takahashi Y 2000 *J. Phys. Soc. Japan* **69** 219
[8] Koyama T, Nakamura H, Kohara T and Takahashi Y 2010 *J. Phys. Soc. Japan* **79** 093704
[9] Yamada H 1991 *J. Phys.: Condens. Matter* **3** 4115
[10] Takahashi Y and Nakano H 2006 *J. Phys.: Condens. Matter* **18** 521
[11] Goto T and Bartashevich M I, *J. Phys.: Condens. Matter* , **10** (1998) 3625
[12] Michor H, El-Hagary M, Mea M Della, Pieper M W, Reissner M, Hilscher G, Khmelevskyi S, Mohn P, Schneider G, Giester G and Rogl P 2004 *Phys. Rev.* B **69** 081404(R)

Correlation between the superconducting pairing symmetry and magnetic anisotropy in f-electron unconventional superconductors

Shinsaku Kambe, Hironori Sakai, Yo Tokunaga and Koji Kaneko[1]

Advanced Science Research Center, Japan Atomic Energy Agency, Tokai-mura, Ibaraki 319-1195, Japan

[1] Quantum Beam Science Directorate, Japan Atomic Energy Agency, Tokai-mura, Ibaraki 319-1195, Japan

E-mail: kambe.shinsaku@jaea.go.jp

Abstract. The superconducting pairing symmetry and the magnetic anisotropy of the normal state are found empirically to be strongly correlated in f-electron unconventional superconductors having crystallographic symmetry lower than cubic. Effectively, there are three categories: 1) In antiferromagnetic systems, unconventional superconductivity appears with singlet (d-wave) pairing or mixed-parity pairing (in non-centrosymmetric compounds) for cases of XY anisotropy. 2) In ferromagnetic systems, unconventional superconductivity with triplet (*e.g.* p-wave) pairing appears for cases of Ising anisotropy. 3) A few exceptional cases: The origin of the observed correlation will be discussed in terms of the orbital f-electron states.

1. Introduction

In f-electron systems, unconventional superconducting (SC) states with an anisotropic superconducting gap are found to occur owing to strong correlations between the f-electrons. In these superconductors, the SC pairing force is considered to be mediated by magnetic fluctuations. In fact, superconductivity in such cases is often near a magnetic instability, so that magnetic ordering is easily induced through variation of a tuning parameter such as pressure or doping [1].

In compounds with strong electron correlations such as heavy fermion superconductors and high T_c cuprates, d-wave superconductivity is favored for antiferromagnetic correlations, whereas p-wave superconductivity is favored for ferromagnetic correlations. This relation between the SC symmetry and the nature of the magnetic correlations has been interpreted qualitatively with theoretical models [2].

In addition to the type of magnetic correlations, the magnetic anisotropies of the f-electron moments themselves, *i.e.* Ising, XY, and Isotropic symmetries, emerge as a result of strong spin-orbit coupling, in contrast to d-electron systems such as high T_c cuprates that show a nearly isotropic magnetism confirmed by neutron scattering [3]. Monthoux and Lonzarich have pointed out that Ising fluctuations qualitatively favor ferromagnetic p-wave superconductivity, whereas isotropic fluctuations favor antiferromagnetic, d-wave superconductivity. This is because p-wave pairing occurs only through the longitudinal magnetic fluctuation channel, whereas d-wave pairing occurs equally through the longitudinal as well as the two transverse fluctuation channels [4]. However, up to now the relation between magnetic anisotropy and superconducting symmetry has not been addressed, and no concrete theoretical description has been reported.

In our previous work [5, 6], the experimental correlation between *d*-wave superconductivity and XY anisotropy has been discussed for the case of heavy fermion superconductors. Recently, a clear correlation between T_c and the strength of XY anisotropy has also been reported for *d*-wave superconductivity in *f*-electron heavy fermion systems [7, 8].

In the present report, we address this issue in greater detail. We classify a series of unconventional, *f*-electron superconductors in terms of their magnetic anisotropy and the symmetry of their superconducting order parameters, finding results which show a strong correlation. Since magnetic anisotropy is a consequence of the relevant orbital states and the crystal field scheme, the results imply a relation between those properties and the nature of unconventional superconductivity.

2. Empirical correlation

In Table 1, the magnetic anisotropy of the magnetic ions and the superconducting symmetry are presented for a series of unconventional, *f*-electron superconductors. Generally, the unconventional superconductivity appears around the magnetic instability point (*i.e.* quantum critical point) in *f*-electron systems. In compounds with ordered ground state at P=0, systems approach to the magnetic instability point under pressure. The magnetic anisotropy at correlation wave vector q_{cor} can be determined experimentally by means of static magnetic susceptibility data for ferromagnetic cases ($q_{cor} = 0$), and with NMR and neutron scattering data for antiferromagnetic cases ($q_{cor} = Q \neq 0$).

For the hexagonal, trigonal, and tetragonal compounds in Table 1, the magnetic anisotropy has axial symmetry at magnetic ions; thus, there are two magnetic symmetry axes in the paramagnetic state, *i.e.* the a-axis in the basal plane and the c-axis perpendicular to the basal plane.

In compounds that do not order magnetically, the magnetic anisotropy is determined by the anisotropy of magnetic fluctuations, *i.e.* the magnetic fluctuation energy along the a-axis Γ_a is smaller than that along the c-axis Γ_c for the XY-type, whereas $\Gamma_a > \Gamma_c$ for the Ising-type. Compounds that order ferromagnetically or antiferromagnetically are designated XY-type if the ordered magnetic moment is in the basal plane and Ising-type if it lies along the c-axis.

Ising-type behavior is assigned to orthorhombic ferromagnetic compounds UGe_2, URhGe and UCoGe, all of which have three magnetic a,b,c-axes at the magnetic ions.

In some compounds, especially compounds showing superconductivity under high pressure, the symmetry of the superconducting order parameter has not yet been determined. Even in these cases, the correlation between the magnetic anisotropy and the type of magnetism (antiferromagnetic or ferromagnetic) is clear. It should be noted that such a correlation has not yet been confirmed in $CeCu_2Si_2$ or $CeIrSi_3$.

Generally, we find there to be three categories of behavior:

- In antiferromagnetic systems, superconductivity occurs for XY behavior with:
  a) singlet pairing (*d*-wave symmetry);
  b) possibly mixed parity pairing for non-centrosymmetric compounds.
- In ferromagnetic systems, superconductivity occurs for Ising behavior with triplet pairing (*e.g.* *p*-wave) symmetry.
- Exceptional cases which cannot be categorized in the above pattern.

2.1. Ordinary cases

The observed strong correlation between the symmetry and the magnetic anisotropy suggests that unconventional superconductivity should be related to the magnetic anisotropy. We call the

Journal of Physics: Conference Series **344** (2012) 012003 doi:10.1088/1742-6596/344/1/012003

Table 1. Magnetic Anisotropy and Symmetry of Superconducting State in Heavy Fermion Systems.

Compound* (Crystal structure)	Ground state(P=0)* (SC T_c) [T_N, T_{HD}]	Magnetic anisotropy at q_{cor} (P=0)	SC symmetry	Anisotropy of static χ(200K)
Ferromagnetic		$\boldsymbol{q_{cor} = 0}$		
UGe_2 (O)	F (0.7K at 1.2GPa)	I ($m_0 \parallel$ [010])[28]	Tri[29]	I[28]
URhGe (O)	SC (0.25K)	I ($\parallel$ [001]) [30]	Tri[31]	I[30]
UCoGe (O)	F+SC (0.6K)	I ($m_0 \parallel$ [001])[32, 33]	?	I[32]
Antiferromagnetic		$\boldsymbol{q_{cor} = Q}$		
$CeCoIn_5$ (T)	SC (2.3K)	XY [34]	d[35]	I[34]
$CeIrIn_5$ (T)	SC (0.4K)	XY [36]	d[37]	I[36]
$CeRhIn_5$ (T)	AF(2K at 2GPa)	XY ($m_Q \parallel$ [110])[23]	Node[17]	I[21]
$PuRuGa_5$ (T)	SC (9K)	XY [5]	d[38]	I(~Iso) [39]
$PuCoGa_5$ (T)	SC (18K)	XY [7]	d[40]	?
$NpPd_5Al_2$ (T)	SC (5K)	XY [41]	d[42]	XY[43]
$CeCu_2Si_2$ (T)	SC (0.6K: S phase)	XY [44]	d[45]	I[46]
$CePd_2Si_2$ (T)	AF (0.4K at 2.8GPa)	XY ($m_Q \parallel$ [110])[47]	?	XY[47]
$CeNi_2Ge_2$ (T)	P (0.4K at 2.6GPa)	XY ($m_Q \parallel$ [110])[48]	?	I[49]
UPd_2Al_3 (H)	AF+SC (1.8K)	XY($m_Q \parallel$ [$10\bar{1}0$])[50]	d[51]	XY[52]
non-centrosymmetric				
$CePt_3Si$ (T NC)	AF+SC (0.75K)	XY ($m_Q \perp$ [001])[53]	Mixed[54]	I[55]
$CeIrSi_3$ (T NC)	AF (1.6K at 2.6GPa)	?	?	XY[56]
$CeRhSi_3$ (T NC)	AF (1K at 2.5GPa)	XY ($m_Q \perp$ [001])[57]	?	XY[58]
Exceptional case				
URu_2Si_2 (T)	SC (1.7K) HD [17K]	AF I ($m_Q \parallel$ [001])$^{1)}$[59]	Node[60]	I[61]
UPt_3 (Trigonal)	SC (0.5K) AF [5.5K]	XY($m_Q \parallel$ [$10\bar{1}0$]) [62]	Tri[63]	XY[64]
$CeRh_2Si_2$ (T)	AF (0.4K at 1.1GPa)	I ($m_Q \parallel$ [001])[13]	?	I[65]
$CeCu_2Ge_2$ (T)	AF (0.6K at 13GPa)$^{2)}$	~Iso($m_Q \parallel\sim$ [$11\frac{1}{2}$]) [16]	?	I[66]
$CeCoGe_3$ (T NC)	AF (0.7K at 6.5GPa)	I ($m_Q \parallel$ [001])[14]	?	I[67]
UNi_2Al_3 (H)	AF+SC (1.2K)	XY($m_Q \parallel$ [$10\bar{1}0$]) [68]	p[69]	XY[70]
Cubic systems				
$CeIn_3$	AF (0.2K at 2.4GPa)	XY? ($m_Q \parallel$ [110])[18]	Node[71]	Iso
$PrOs_4Sb_{12}$	SC (1.85K)	?	p?[72]	Iso
UBe_{13}	SC (0.9K)$^{3)}$	?	p?[73]	Iso

T: Tetragonal, H: Hexagonal, O: Orthorhombic, NC: Non-centrosymmetric, AF: Antiferromagnetic ordering, F: Ferromagnetic ordering, P: Paramagnetic, HD: Hidden ordering, SC: Superconducting, I: Ising, Iso: Isotropic, m_Q: Antiferromagnetic ordered moment, m_0: Ferromagnetic ordered moment, Node: Existence of SC gap node is suggested, Tri: Triplet pairing, Mixed: Mixed parity paring,
* Please see ref. [27]. 1) Antiferromagnetic ordered state under pressure, 2) There is another maximum of T_c =2K at 17 GPa, 3) There may be another SC phase with T_c = 0.75K.

Journal of Physics: Conference Series **344** (2012) 012003 doi:10.1088/1742-6596/344/1/012003

first and second categories "*ordinary cases*", since it appears natural that the antiferromagnetic-XY and ferromagnetic-Ising systems are found to be more favourable for *d*-wave and *p*-wave superconductivity, respectively, in compounds with symmetry lower than cubic. It should be noted, however, that *d*-wave superconductivity occurs in systems with strong XY anisotropy rather than in nearly isotropic systems in the case of f-electron compounds [5, 7], an effect which goes beyond the previous theoretical explanation [4].

The possibility that mixed symmetry occurs in compounds without inversion symmetry is well known by now.

2.2. *Exceptional case*

Now let us examine exceptional cases which do not show the same pattern as the ordinary cases. In a certain sense, these exceptional cases are more curious compared with ordinary cases, since there may be interesting reasons for the exception.

In URu_2Si_2, superconductivity appears only in the phase with hidden order, thus the Ising antiferromagnetic fluctuations found near the antiferromagnetic ordered phase may not be relevant to the superconductivity. In fact, an incommensurate magnetic excitation at $q_{cor} = (1 \pm 0.4, 0, 0)$ was found recently in the hidden ordered phase [9], which is different from $q_{cor} = (0, 0, 1)$, which characterizes the antiferromagnetic ordering. The superconductivity could be mediated by fluctuations of the hidden order parameter, which may be multipolar (spin$\otimes$orbital) ordering [10, 11].

In UPt_3, the nature of the antiferromagnetic ordering is quite different from an ordinary case, *i.e.* appears to be a tiny moment ($\sim 0.02\mu_B$) which exhibits slow fluctuations [12]. Thus, the relation between superconductivity and antiferromagnetic fluctuations may be different from usual. In fact, it has been proposed that superconductivity is mediated by other fluctuations (*e.g.* ferromagnetic fluctuations) in this compound [12].

In $CeRh_2Si_2$, $CeCoGe_3$ and $CeCu_2Ge_2$, superconductivity occurs in Ising or Isotropic-like antiferromagnets. Remarkably, these compounds have a complex antiferromagnetic ordered state in common. In $CeRh_2Si_2$, there are two antiferromagnetically ordered phases with different multi-$\vec{k}$ structures, *i.e.* $4\vec{k}$-structure below 36 K and $2\vec{k}$-structure below 25 K [13], although the magnetic structure around the critical pressure 1.1GPa for magnetic instability has not yet been determined. In $CeCoGe_3$ there are three different antiferromagnetically ordered phases at $P = 0$, occurring at T_{N1}=20 K, T_{N2}=11.5 K, and T_{N3}=7.5 K [14], in addition, the transitions at T_{N2} and T_{N3} may be 1st order phase transitions with a change of ordered structure [15]. In $CeCu_2Ge_2$, two different antiferromagnetically ordered states appear on substitution of a few percent of Ni in the place of Cu [16]. In this compound, a complex incommensurate [$q_{cor} = (0.284, 0.284, 0.543)$] ordered structure with a sinusoidally modulated amplitude is found with m_Q inclined 10 degree from q_{cor}.

Perhaps the exceptional behavior in these compounds is related to these complex ordered phases, indicating a frustrated exchange interaction. In any case, in order to confirm these exceptional properties, a determination of magnetic anisotropy and superconducting symmetry under pressure is certainly necessary for these compounds. In fact, q_{cor} may be found to depend on pressure [17].

It is somewhat surprising that UNi_2Al_3 and UPd_2Al_3 have a different superconducting symmetry, since these compounds are quite similar. As it was difficult to synthesize a high quality sample of UNi_2Al_3, it may be necessary to check the superconducting symmetry of UNi_2Al_3 using a better sample.

2.3. *Cubic case*

Finally, we address the case of cubic systems. In the cubic compound $CeIn_3$ with cubic local symmetry for Ce, magnetic fluctuations should be isotropic in the paramagnetic state. However, a recent study implied that the AF ordered moment is along the [110] direction [18], indicating that XY anisotropy develops near the ordered state. It may then be possible to categorize $CeIn_3$ as an ordinary case for future work.

Concerning the systems UBe_{13} and $PrOs_4Sb_{12}$, it is difficult to discuss the pairing symmetry and magnetism here, since many physical properties remain controversial at the present time.

3. Orbital state and superconductivity

Several interesting issues emerge from the categorizations we propose here.

- Is XY more favourable than Ising-type fluctuation for unconventional superconductivity in f-electron antiferromagnets?
- Do anisotropic fluctuations develop around the superconducting transition even in cubic systems?
- What is the origin of the exceptional cases?

It is clear that the magnetic anisotropy is due to the spin-orbit coupling, crystal field and anisotropic orbital states. Thus, the relation between the orbital states and the superconductivity needs to be considered. For example, a particular orbital state with XY-anisotropy can be more favourable than an isotropic one for antiferromagnetic cases. Concerning fluctuations relevant to superconductivity, orbital and multipolar fluctuations should be addressed in addition to spin fluctuations. Although the orbital states are difficult to treat in itinerant systems, the following theoretical approaches have recently been proposed.

Takimoto *et al* [19] have pointed out that an orbital splitting energy Δ between the $\Gamma_7^{(1)}$ and $\Gamma_7^{(2)}$ levels can be relevant to T_c in tetragonal Ce-based superconductors. In the crystal field scheme for the Ce^{3+} ion in tetragonal symmetry, two $\Gamma_7^{(1,2)}$ and one Γ_6 doublets are formed. In $CeTIn_5$ superconductors for example, the following $\Gamma_7^{(1,2)}$ level is lower than the Γ_6.

$$
\begin{aligned}
\mid \Gamma_{6\pm}\rangle &= \mid \pm\frac{1}{2}\rangle \\
\mid \Gamma_{7\pm}^{(1)}\rangle &= \sqrt{\frac{5}{6}} \mid \pm\frac{5}{2}\rangle + \sqrt{\frac{1}{6}} \mid \mp\frac{3}{2}\rangle \\
\mid \Gamma_{7\pm}^{(2)}\rangle &= \sqrt{\frac{1}{6}} \mid \pm\frac{5}{2}\rangle - \sqrt{\frac{5}{6}} \mid \mp\frac{3}{2}\rangle
\end{aligned}
\tag{1}
$$

In this model, the T_c for d-wave superconductivity increases with increasing Δ up to a finite value, after which an antiferromagnetically ordered state appears. The enhancement of T_c is ascribed to enhancement of spin fluctuations and suppression of orbital fluctuations which occur with increasing Δ. If the antiferromagnetic XY anisotropy is correlated with Δ, this model may explain the correlations between d-wave superconductivity and antiferromagnetic XY anisotropy through the spin and orbital fluctuations.

Flint *et al* [20] have proposed a composite superconducting pairing mechanism for actinide superconductors. In this model, Cooper pairs are formed using three different orbital states. This model could potentially identify the relation between the superconductivity and the magnetic anisotropy.

In f-electron systems, there may be either a localized or an itinerant origin for the observed magnetic anisotropy: 1) Local magnetic anisotropy determined by the crystal field and the f-electron state; 2) the Ruderman, Kittel, Kasuya and Yosida (RKKY) interaction due to hybridization between f-electrons and conduction electrons.

If the localised character is strong, the anisotropy of the static susceptibility can be qualitatively explained in terms of a crystal field scheme based on the LS-coupling description.

This is the case for many Ce compounds [21]. However, as shown in Table 1, the anisotropy of the static susceptibility ($q = 0$) does not always coincide with the magnetic anisotropy at the antiferromagnetic wave vector ($q = Q$). This deviation indicates that the RKKY exchange interaction has a q-dependence, an effect which is clearly related to the itinerant nature of the system. In order to treat the actual magnetic anisotropy over a wide range of q-values, it is necessary to treat correctly the intermediate properties of f-electrons between localised and itinerant. The jj-coupling description may be suitable for this purpose [22].

It should be noted that the dimensionality and the magnetic anisotropy should be distinguished. Strong XY anisotropy does not mean strong two dimensional magnetic character for systems. The measure of two dimensionality is the anisotropy of the magnetic correlation length ξ_a/ξ_c (in units of the lattice parameter ratio a/c), which is typically $0.5 \sim 5$ in the tetragonal and hexagonal unconventional superconductors in Table 1 ($\xi_a/\xi_c = \infty$ in the two dimensional systems) [23, 24]. This fact indicates that these compounds are anisotropic, but still three dimensional.

Recently, a possible valence (monopolar) fluctuation-mediated superconductivity has been discussed [25]. The orbital character can be modified at a valence transition. Clearly, valence fluctuations are an important aspect of f-electron systems; however, the relation between this scenario and the present findings is an open question at the present stage of things. The valence and magnetic fluctuations can also be mixed, a situation which may be realized in $CeCu_2Ge_2$ and $CeCu_2Si_2$ [26]

4. Conclusion

The symmetry character of unconventional superconductivity in f-electron systems is correlated with the magnetic anisotropy in the normal state. This indicates strongly that the spin-orbit coupling and orbital character must be incorporated in the mechanism of unconventional superconductivity in f-electron systems. If spin, orbital, and multipolar degrees of freedom are correctly treated in the mechanism of superconductivity, the existence of these categories is expected to be elucidated in a unified picture.

Acknowledgments

We acknowledge stimulating discussions with T. Takimoto, J. Flouquet, T. D. Matsuda, Y. Haga, R.E. Walstedt and K. Kubo. This work was supported by a Grant-in-Aid for Scientific Research on Innovative Areas "Heavy Electrons" of The Ministry of Education, Culture, Sports, Science, and Technology, Japan.

The authors thank the Yukawa Institute for Theoretical Physics at Kyoto University, where this work was initiated during the YITP-W-10-12 on "International and Interdisciplinary Workshop on Novel Phenomena in Integrated Complex Sciences: from Non-living to Living Systems". This workshop was supported in part by the Grant-in-Aid for the Global COE Program "The Next Generation of Physics, Spun from Universality and Emergence" from the Ministry of Education, Culture, Sports, Science and Technology (MEXT) of Japan.

References

[1] Flouquet J 2005 *Prog. Low Temp. Phys.* **XV** (Elsevier) p 3
[2] Miyake K, Shimitt-Rink S and Varma C M 1986 *Phys. Rev. B* **34** 6554
[3] Kastner M A, Birgeneau R J, Shirane G and Endoh Y 1988 *Rev. Mod. Phys.* **70** 897
[4] Monthoux P and Lonzarich G G 2001 *Phys. Rev. B* **63** 054529
[5] Kambe S, Tokunaga Y, Sakai H, Chudo H, Haga Y, Matsuda T D, Walstedt R E 2010 *Phys. Rev. B* **75** 140509R
[6] Walstedt R E, Kambe S, Tokunaga Y, Sakai H 2007 *J. Phys. Soc. Jpn.* **76** 082001
[7] Baek S H, Sakai H, Bauer E D, Mitchell J N, Kennison J A , Ronning F and Thompson J D 2010 *Phys. Rev. Lett.* **105** 217002

[8] Sakai H, Kambe S, Tokunaga Y, Haga Y, Baek S H, Ronning F, Bauer E D and Thompson J D 2010 *Mater. Res. Soc. symp. Proc.* **1264** Z12-01
[9] Broholm C, Kjems J K, Buyers W J L, Matthews P, Palstra T T M, Menovsky A A and Mydosh J A 1987 *Phys. Rev. Lett.* **58** 1467
[10] Harima H, Miyake K and Flouquet J 2010 *J. Phys. Soc. Jpn.* **79** 033705
[11] Haule K and Kotliar G 2009 *Nat. Phys.* **5** 796
[12] van Dijk N H, Rodière P, Fak B, Huxley A and Flouquet J 2002 *Physica B* **319** 220
[13] Kawarazaki S, Sato M, Miyako Y, Chigusa N, Watanabe K, Metoki N, Koike Y and Nishi M 2000 *Phys. Rev. B* **61** 4167
[14] Kaneko K, Metoki N, Takeuchi T, Matsuda T D, Haga Y, Thamizhavel A, Settai R and Ōnuki Y 2009 *J. Phys. Conf. Series* **150** 042082
[15] Kaneko K *et al* unpublished.
[16] Krimmel A, Loidl A, Schober H and Canfield P C 1997 *Phys. Rev. B* **55** 6416
[17] Yashima M, Kawasaki S, Mukuda H, Kitaoka Y, Shishido H, Settai R, and Ōnuki Y 2007 *Phys. Rev. B* **76** 20509R
[18] Demuynck S, Sandratskii L, Cottenier S, Meersschaut J and Rots M 2000 *J. Phys. Condens. Matter* **12** 4629
[19] Takimoto T, Hotta T and Ueda K 2004 *Phys. Rev. B* **69**, 104504
[20] Flint R, Dzero M and Coleman P 2008 *Nat. Phys.* **4** 643
[21] Takeuchi T, Inoue T, Sugiyama K, Aoki D, Tokiwa Y, Haga Y, Kindo K and Ōnuki Y 2001 *J. Phys. Soc. Jpn.* **70** 877
[22] Hotta T and Ueda K 2003 *Phys. Rev. B* **67** 104518
[23] Bao W, Pagliuso P G, Sarrao J L, Thompson J D, Fisk Z, Lynn J W and Erwin R W 2000 *Phys. Rev. B* **62** 14621
[24] Sato N, Aso N, Hirota K, Komatsubara T, Endoh Y, Shapiro S M, Lander G H and Kakurai K 1996 *Phys. Rev. B* **53** 14043
[25] Watanabe S and Miyake K 2010 *Phys. Rev. Lett.* **105** 186403
[26] Fujiwara K, Okazaki Y, Hata Y, Miyoshi K, Takeuchi J, Kobayashi T C, Geibel C and Steglich F 2010 *Physica Stat. Soli.* **247** 608
[27] Pfleiderer C 2009 *Rev. Mod. Phys.* **81** 1551 and references therein
[28] Ōnuki Y, Ukon I, Yun S W, Umehara I, Satoh K, Fukuhara T, Sato H, Takayanagi S, Shikama M and Ochiai A 1992 *J. Phys. Soc. Jpn.* **61** 293
[29] Huxley A, Sheikin I, Ressouche E, Kernavanois N, Braithwaite D, Calemczuk R and Flouquet J 2001 *Phys. Rev. B* **63** 144519
[30] Proke K, Tahara T, Echizen Y, Takabatake T, Fujita T, Hagmusa I H, Klaasse J C P, Bruck E, de Boer F R, Divi M and Sechovsky V 2002 *Physica B* **311** 220
[31] Hardy F and Huxley A D 2005 *Phys. Rev. Lett.* **94** 247006
[32] Huy N T, de Nijs D E, Huang Y K and de Visser A 2008 *Phys. Rev. Lett.* **100** 77002
[33] Ihara Y, Hattori T, Ishida K, Nakai Y, Osaki E, Deguchi K, Sato N K, and Satoh I 2010 *Phys. Rev. Lett.* **105** 206403
[34] Curro N J and Pines D 2007 *J. Phys. Chem. Solids* **68**, 2028
[35] Izawa K, Yamaguchi H, Matsuda Y, Shishido H, Settai R and Ōnuki Y 2001 *Phys. Rev. Lett.* **87** 057002
[36] Kambe S, Tokunaga Y, Sakai H, Chudo H, Haga Y, Matsuda T D, and Walstedt R E 2010 *Phys. Rev. B* **81** 140405R
[37] Zheng G q, Tanabe K, Mito T, Kawasaki S, Kitaoka Y, Aoki D, Haga Y and Ōnuki Y 2001 *Phys. Rev. Lett.* **86** 4664
[38] Sakai H, Tokunga Y, Fujimoto T, Kambe S, Walstedt R E, Yasuoka H, Aoki D, Homma Y, Yamamoto E, Nakamura A, Shiokawa Y, Nakajima K, Arai Y, Matsuda T D, Haga Y and Ōnuki Y 2005 *J. Phys. Soc. Jpn.* **74** 1710
[39] Haga Y, Aoki D, Matsuda T D, Nakajima K, Arai Y, Yamamoto E, Nakamura A, Homma Y, Shiokawa Y and Ōnuki Y 2005 *J. Phys. Soc. Jpn.* **74** 1698
[40] Curro N J, Caldwell T, Bauer E D, Morales L A, Graf M J, Bang Y, Balatsky A V, Thompson J D and Sarrao J L 2005 *Nature* **434** 622
[41] Chudo H, Sakai H, Tokunaga Y, Kambe S, Aoki D, Homma Y, Haga Y, Matsuda T D, Ōnuki Y and Yasuoka H 2010 *J. Phys. Soc. Jpn.* **79** 053704
[42] Chudo H, Sakai H, Tokunaga Y, Kambe S, Aoki D, Homma Y, Shiokawa Y, Haga Y, Ikeda S, Matsuda T D, Ōnuki Y and Yasuoka H 2008 *J. Phys. Soc. Jpn.* **77** 083702
[43] Aoki D, Haga Y, Matsuda T D, Tateiwa N, Ikeda S, Homma Y, Sakai H, Shiokawa Y, Yamamoto E, Nakamura A, Settai R and Ōnuki Y 2007 *J. Phys. Soc. Jpn.* **76** 083708
[44] O. Stockert private communications

[45] Ishida K, Kawasaki Y, Tabuchi K, Kashima K, Kitaoka Y, Asayama K, Geibel C and Steglich F 1999 *Phys. Rev. Lett.* **82** 5353
[46] Ōnuki Y, Hirai T, Komatsubara T, Takayanagi S, Sumiyama A, Furukawa A, Oda Y and Nagano H 1985 *J. Magn. Magn. Mater.* **52** 338
[47] van Dijk N H, Fak B, Charvolin T, Lejay P and Mignot J M 2000 *Phys. Rev. B* **61** 8922
[48] Fak B, Flouquet J,Lapertot G, Fukuhara T and Kadowaki H 2000 *J. Phys. Condens. Matter* **12** 5423
[49] Fukuhara T, Maezawa K, Ohkuni H, Sakurai J, Sato H, Azuma H, Sugiyama K, Ōnuki Y and Kindo K 1996 *J. Phys. Soc. Jpn.* **65** 1559
[50] Krimmel A, Fischer P, Roessli B, Maletta H, Geibel C, Schank C, Grauel A, Loidl A and Stegiich F 1992 *Z. Phys. B - Condensed Matter* **86**, 161
[51] Watanabe T, Izawa K, Kasahara Y, Haga Y, Onuki Y, Thalmeier P, Maki K and Matsuda Y 2004 *Phys. Rev. B* **70** 184502
[52] Grauel A, Bohm A, Geibel C, Weber G, Steglich F, Sato N, Komatsubara T and Horn S 1993 *J. App. Phys.* **73** 5421
[53] Metoki N, Kaneko K, Matsuda T D, Galatanu A, Takeuchi T, Hashimoto S, Ueda T, Settai R, Onuki Y and Bernhoeft N 2004 *J. Phys. Condens. Matter.* **16** L207
[54] Yogi M, Mukuda H, Kitaoka Y, Hashimoto S, Yasuda T, Settai R, Matsuda T D, Haga Y, Ōnuki Y, Rogl P and Bauer E 2006 *J. Phys. Soc. Jpn.* **75** 13709
[55] Takeuchi T, Hashimoto S, Yasuda T, Shishido H, Ueda T, Yamada M, Obiraki Y, Shiimoto M, Kohara H, Yamamoto T, Sugiyama K, Kindo K, Matsuda T D, Haga Y, Aoki Y, Sato H, Settai R and Ōnuki Y 2004 *J. Phys. Condens. Matter.* **16** L333
[56] Sugitani I, Okuda Y, Shishido H, Yamada T, Thamizhavel A, Yamamoto E, Matsuda T D, Haga Y, Takeuchi T, Settai R and Ōnuki Y 2006 *J. Phys. Soc. Jpn.* **75** 43703
[57] Aso N *et al* private communications
[58] Muro Y, Ishikawa M, Hirota K, Hiroi Z, Takeda N, Kimura N, and Aoki H 2007 *J. Phys. Soc. Jpn.* **76** 33706
[59] Amitsuka H, Sato M, Metoki N, Yokoyama M, Kuwahara K, Sakakibara T, Morimoto H, Kawarazaki S, Miyako Y, and Mydosh J A 1999 *Phys. Rev. Lett.* **83** 5114
[60] Matsuda K, Kohori Y and Kohara T 1996 *J. Phys. Soc. Jpn.* **65** 679
[61] Palstra T T M, Menovsky A A, van den Berg J, Dirkmaat A J, Kes P H, Nieuwenhuys G J and Mydosh J A 1985 *Phys. Rev. Lett.* **55** 2727
[62] Hayden S M, Taillefer L, Vettier C, Flouquet J 1992 *Phys. Rev. B* **46** 8675
[63] Tou H, Kitaoka Y, Ishida K, Asayama K, Kimura N, Onuki Y, Yamamoto E, Haga Y and Maezawa K 1998 *Phys. Rev. Lett.* **80** 3129
[64] Frings P H, Franse J J M , de Boer F R and Menovsky A 1983 *J. Magn. Magn. Mater.* **31-34** 240
[65] Abe H, Kitazawaa H, Suzukia H, Kido G and Matsumoto T 1998 *J. Magn. Magn. Mater.* **177-181** 479
[66] Matsuda T D *et al* private communications
[67] Thamizhavel A, Takeuchi T, Matsuda T D, Haga Y, Sugiyama K, Settai R and Onuki Y 2005 *J. Phys. Soc. Jpn.* **74** 1858
[68] Lussier J G, Mao M, Schröder A, Garrett J D and Gaulin B D, Shapiro S M and Buyers W J L 1997 *Phys. Rev. B* **56** 11749
[69] Ishida K, Ozaki D, Kamatsuka T, Tou H, Kyogaku M, Kitaoka Y, Tateiwa N, Sato N K, Aso N, Geibel C and Steglich F 2002 *Phys. Rev. Lett.* **89** 37002
[70] Sato N, Koga N and Komatsubara T 1996 *J. Phys. Soc. Jpn.* **65** 1555
[71] Kawasaki S, Mito T, Kawasaki Y, Zheng G q, Kitaoka Y, Shishido H, Araki S, Settai R and Ōnuki Y 2002 *Phys. Rev. B* **66** 54521
[72] Higemoto W, Saha S R, Koda A, Ohishi K, Kadono R, Aoki Y, Sugawara H and Sato H 2007 *Phys. Rev. B* **75** 020510
[73] Ott H R, Rudigier H, Rice T M, Ueda K, Fisk Z and Smith J L 1984 *Phys. Rev. Lett.* **52** 1915

IIW-NPICS 2010 IOP Publishing
Journal of Physics: Conference Series **344** (2012) 012004 doi:10.1088/1742-6596/344/1/012004

Thinking Patterns, Brain Activity and Strategy Choice

Kazuo Nishimura[1], Akira Okada[2], Michiyo Inagawa[3] and Yoshikazu Tobinaga[4]
[1]Institute of Economic Research, Kyoto University
Yoshida-honmachi, Sakyo-ku, Kyoto 606-8501, JAPAN
E-mail: nishimura@kier.kyoto-u.ac.jp

[2] Department of Economics, Hitotsubashi University.

[3] Graduate School of Education, Kyoto University.

[4] Elegaphy, Inc.,

Abstract: In this study we analyzed the relationship between thinking patterns, behavior and associated brain activity. Subjects completed a self-report assessing whether they could voluntarily stop thinking or not, and were then divided into two groups: those with the ability to stop thinking and those without. We measured subjects' brain activity using magnetoencephalography while giving them a series of tasks intended to encourage or discourage spontaneous thinking. Our findings revealed differences between the two groups in terms of which portions of the brain were active during the two types of task. A second questionnaire confirmed a relationship between the ability to stop thinking and strategy choices in a dilemma game. We found that subjects without the ability to stop thinking had a tendency to choose cooperative behavior.

1. Introduction

Economics has attached importance to studies focusing on the allocation of resources in societies made up of heterogeneous agents [1]. It is also a common practice in economics to group these agents into risk lover and risk averter categories according to their attitudes toward risks [2]. Game theory has studied the interdependent relationships among heterogeneous agents, for instance with regard to conflict of interest and cooperation, as well as these agents' decision-making strategies [3],[4],[5].

Neuroeconomics has recently received growing attention and has been the subject of an increasing number of articles in scientific journals. This field explores the relationship between economic behavior and associated brain activity and aims to elucidate the ways in which people make decisions. In this sense, it may be described as the synthesis of brain science research and experimental and behavioral economics. More precisely, experiments in this field of study involve performing brain activity scans using fMRI, PET, MEG, etc. These functional neuroimaging devices and measurements of the time courses of different brain events provide us with records of the activities of neurons and neurotransmitters as people engage in economic decision making such as choosing between risks and rewards and interacting with other people. Meta-analyses of functional imaging studies have shown that the prefrontal cortex of the brain operates on choices to be faced in the distant future, while the limbic system works on choices to be encountered in the near future [6]. Thus, these types of study may be regarded as an effective approach to understanding individuals' attitudes and behaviors.

Despite the fact that researchers have recognized the importance of focusing expressly on individual differences, the scope of studies on economic behaviors of heterogeneous agents has been limited. Indeed, there has been almost no research on how differences in individual thinking patterns affect decision making. We therefore aim to study how the way in which one thinks affects his or her behavior.

Accordingly, this study focuses on the reasons for differences in decision making among heterogeneous agents as well as differences in brain activity involved in the decision-making process. Few studies of this sort have been performed thus far, but we believe that this approach to investigating thinking traits and decision making will lay the foundation for elucidating the way in which consumers and investors decide on economic issues and future choices. In our study, we divided our subjects into two types based on their thinking patterns: Group A consisted of individuals who claimed they could stop themselves from thinking, whereas Group B consisted of individuals who reported that they could not. Section 2 examines the survey results in light of recent findings concerning the relationship between the ability to stop thinking and brain activity.

Section 3 reports the results of a questionnaire that was administered to examine the relationship between thinking patterns and decision making in a dilemma game. In this game, subjects were asked to choose between cooperative and non-cooperative behavior. The survey results showed that subjects without the ability to stop thinking had a tendency to choose cooperative behavior.

2. Thinking patterns – can one stop thinking or not?

2.1. Materials and methods

Data from 14 healthy subjects (age: 17–34) were analyzed. We recruited participants from among our colleagues and their families.

2.1.1. Experimental paradigm. We previously conducted studies to measure the brain activity underlying human thought [7], [8], [9]. To do this, we prepared tasks that would direct subjects to "think" or "stop thinking," and then measured these subjects' brain activities using magnetoencephalography (MEG).

In these experiments, subjects lay down in a neuroimaging device and were assigned tasks designed to encourage or discourage thinking. For instance, subjects were asked to recall images or engage in linguistic activity, and were also requested to stop thinking for 20 sec. In order to verify our previous results, we used the same tasks in this study.

Our test protocol asked subjects to picture images of (1) Kiyomizu Temple and (2) the Diet Building, (3) to recall the Chinese zodiac, (4) to recall a conversation they had earlier in the day, and (5) and (6) to not think at all (Table 1). We asked subjects to go through this series of tasks twice.

Tasks 1 and 2 stimulated subjects' thinking by encouraging their recall of a familiar place, while Tasks 3 and 4 tested their ability to recall words. The final tasks (5 and 6) were intended to examine subjects' capability to suspend their thinking. Our goal was to determine whether we could detect differences in brain activity between Tasks 1–4 and Tasks 5–6.

We used a 100-msec beep tone as the signal for the subjects to advance to the next task. During the tests, brain activity was monitored as changes in the spontaneous brain magnetic field.

Data acquisition began at the first beep, though the data we actually used began 500 msec afterward. Thus, we sampled only the brain's thinking activity, not the auditory-evoked response elicited by the beep. Measurements were taken every 50 msec.

Table 1. "Thinking" Tasks during MEG

Proceed from Task 1 to Task 6. Advance to the next task at the beep signals. If you finish Task 6, begin again with Task 1.

Task	What to do:
1	Picture Kiyomizu Temple
2	Picture the Diet Building
3	Think of the name of each animal in the Chinese Zodiac
4	Recall a conversation you had earlier today

Sit still, relax, and try not to think at all.	
5	Rest (eyes open) and do not think at all
6	Rest (eyes open) and do not think at all

2.1.2. Data acquisition. Assessing higher cognitive functions such as thought in terms of brain activity can only be accomplished using methods that measure intracerebral bloodstream or cranial nerve currents. When frequency discrimination is required, as in this experiment, measurement of cranial nerve currents is more suitable, as it allows for accurate detection of time series signals.

In order to assess higher functions such as cognition, we employed MEG, which can detect ultra-low intensity cerebral magnetic fields at the femto-Tesla (fT) level [10].

Actual cerebral measurements are fairly unstable and are liable to include various artifacts. In particular, the measurement of spontaneous cerebral activity such as thought is extremely difficult. To confirm that the electric current dipoles were measured accurately as brain activity, we analyzed the relatively large waveform (an alpha wave) that appeared in the occipital region and then superimposed the electric current dipoles that had been obtained. As a result, we could confirm that the electric current dipoles corresponded to the pattern of the sulci in the brain, which seems to represent brain activity.

2.1.3. Data analysis. Our experiment used a 64-channel whole head-type superconducting quantum interference device (SQUID) made by CTF System Inc., an axial type of neuromagnetometer.

Let $B=(B_x, B_y, B_z)$ be the magnetic field vector. B_z is the element of B along the z-axis that is perpendicular to the head surface. The 64-sensor neuromagnetometer can measure the first derivative of B_z (an element in the z-axis of the magnetic field) through individual SQUID sensors installed on the helmet, or it can calculate $(\partial B_z/\partial z)_i$ based on the data from SQUID sensor *i*. Its dimension is $fT/cm(Hz)^{1/2}$, and relative to the 64 parts of the head surface, *i* varies from 1 to 64. A total of 64 pieces of magnetic field data is recorded at a particular interval and $(\partial B_z/\partial z)_i$ $(i=1, 2, \cdots, 64)$ are calculated.

Single dipole method

Based on Biot–Savart's law, the Single Dipole Method is a common technique used to estimate a signal source of magnetic field distribution that emerges on the surface of the head. This technique was based on the hypothesis that the brain magnetic field does not distort. We therefore surmised that the first-order approximation of reading values would not be affected because the influence of the distribution current, the so-called "volume current," could be balanced out by the spatial symmetry. The equivalent current dipole, displayed using three-dimensional vectors, should subsequently emerge from the brain [11].

Current dipoles can be estimated by solving the inverse problem of the magnetic field distribution as projected on the head surface. For the estimation procedures, we first drew a contour map from a magnetic field in reference to the measured values of $(\partial B_z/\partial z)_i$. This contour map allowed us to estimate the signal source by following the least-squares estimation method. Using this method, the signal source could be defined as in the middle position of the extreme and the sink identified on the magnetic field. The estimated signal source should therefore have been closer to the actual value. We applied the Single Dipole Method only when goodness of fit was sufficiently high to guarantee accuracy in the least-squares estimation.

2.2 Results

We examined the differences in brain activity between subjects who were able to stop thinking and those who were not. The frontal region was activated during thinking tasks whereas occipital activity was noticeable during tasks that required cessation of thinking, though individual differences were also observed (see Fig. 1). In Subject A, frontal region activity diminished slightly but was nonetheless present after transitioning to the stop-thinking tasks. This suggests the possibility that the subject continued thinking when presented with the stop-thinking tasks. On the other hand, Subject O demonstrated a clear concentration of activity in the occipital region during the stop-thinking task, suggesting that the subject did actually stop thinking when asked to do so.

Journal of Physics: Conference Series **344** (2012) 012004 doi:10.1088/1742-6596/344/1/012004

These results indicate that individual differences in thinking methods correspond to the degree with which localized functions are activated in different brain regions. The same pattern of function localization is generally observed among individuals, but the specific way these functions are activated is considered to underlie individual differences in thinking methods.

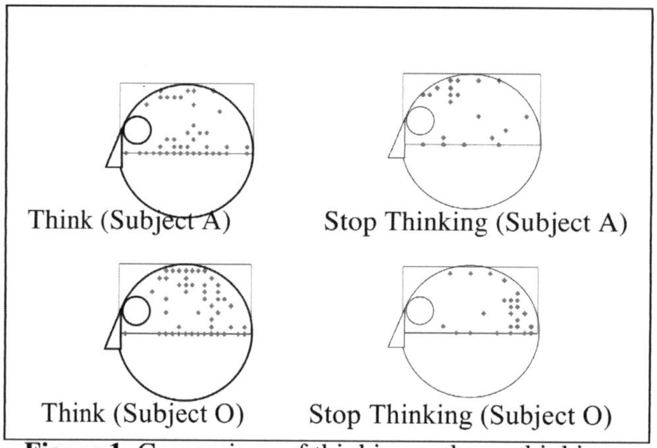

Figure 1. Comparison of thinking and non-thinking states between Subject A and Subject O, showing analysis results of α-wave constituent signals in cerebral magnetic fields (8–12 Hz). The side interior view demonstrates frontal–occipital differences.

3. Game theory – decision making

3.1. Material and method

In an attempt to analyze the relationship between thinking patterns and decision making, a questionnaire (Table 2) was administered to 51 undergraduate and graduate students at the Faculty of Economics, Hitotsubashi University.

We prepared a dichotomous-choice, single-answer questionnaire for each task. The students were asked to perform various tasks and were classified into groups accordingly.

We requested that participants try not to think, and afterwards asked them if they were able to avoid doing so. Thirty-two participants (62.7%) replied that they were able to stop thinking, while 19 (37.3%) responded that they were not.

We evaluated whether there were any behavioral differences between these two groups by assessing strategy choices in a dilemma game as described in the next section.

Table 2. Self-report on thinking.

Please close your eyes and try not to think at all. Use the scale below to evaluate whether or not you were able to stop thinking.

(1,	0,	−1)
stopped	not certain	could not stop

3.2. Experiment – Dilemma game

A dilemma game requires two individuals to choose between two behaviors without consulting each other. An individual's profit (or benefit) depends not only on his or her behavior but also on the other person's behavior, as shown in the following profit matrix.

In Table 3, the two strategy combinations, (A, B) and (B, A), represent the Nash equilibria. If both subjects opt for Strategy A, they both gain a profit of 100. If only one of them opts for Strategy B, the individual who chooses Strategy B gains the game's maximum profit of 200, while the individual who continues with Strategy A receives a profit of only 80. If both individuals opt for Strategy B with the aim of gaining the maximum profit of 200, both are awarded the game's minimum profit of 50. In view of this relationship between behavior and profit, Strategy A can be interpreted as representing "cooperative behavior," and Strategy B as representing "non-cooperative behavior."

This dilemma game has two Nash equilibria, so players cannot easily decide which strategy to opt for. The dilemma is whether to cooperate with the other player or not. These features of the exercise allowed us to evaluate how each player's strategy choice was related to his or her thinking traits, especially his or her ability or inability to stop thinking. The game also required each player to assess the other player's strategy in making decisions, and we included an item in our questionnaire to examine the accuracy with which each individual was able to do so.

Table 3. Dilemma game.

You and the other player will respectively choose between A and B. Depending on the combination of the choices, you will both receive points as shown in the table below. The left-hand figure in each box represents the points you will earn. What do you think the other player will choose? What will you choose?

		Other player	
		A	B
You	A	100, 100	80, 200
	B	200, 80	50, 50

3.3. Results and discussion

We administered the questionnaire shown in Table 3 regarding the strategies employed in the dilemma game in order to verify the relationship between the ability to stop thinking and decision making in a social environment where decision making is required. The subjects of this questionnaire were the same Hitotsubashi University students who had studied game theory and had participated in Experiment 1. However, two participants were removed from the sample because they did not answer this experiment's questionnaire.

The results are shown in Figs. 2 and 3. Figure 2 demonstrates that in the group with the ability to stop thinking, 16 players, or 51.6%, estimated that their opponent would opt for Strategy A, while 15, or 48.4%, predicted that the other player would choose Strategy B. In the group without the ability to stop thinking, nine players anticipated that their opponent would choose Strategy A, while nine predicted a choice of Strategy B, thus also yielding a ratio of 50%. Figure 3 shows that each player's own strategy also differed depending on his or her ability to stop thinking. Sixteen (51.6%) members of the group with the ability to stop thinking chose the non-cooperative Strategy B, whereas only four individuals (22.2%) without the ability to stop thinking did so.

We compared the ratio differences between groups and found no significant difference between the groups with and without the ability to stop thinking in terms of the ability to predict the opposing player's behavior ($z = -0.10888$, $q = 1.644854$, $P = 0.5433512$, $p < .05$). On the contrary, there was a significant difference between groups in players' own strategy choices ($z = 2.107952$, $q = 1.644854$, $P = 0.021798$, $p < .05$). These results demonstrated that those who were unable to stop thinking were more likely to choose cooperative behavior.

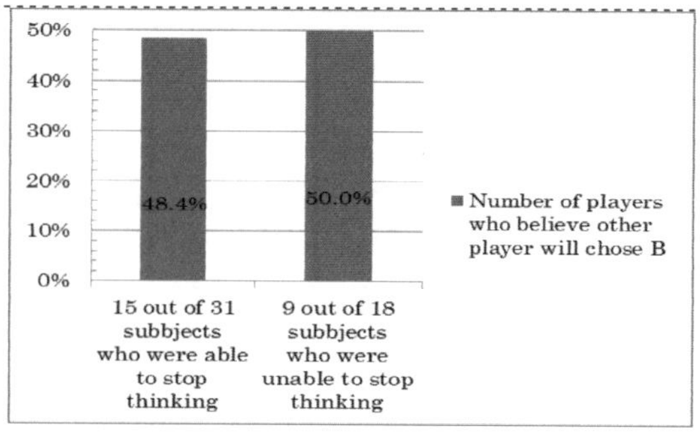

Figure 2. Expectation of opposing player's non-cooperative choice.

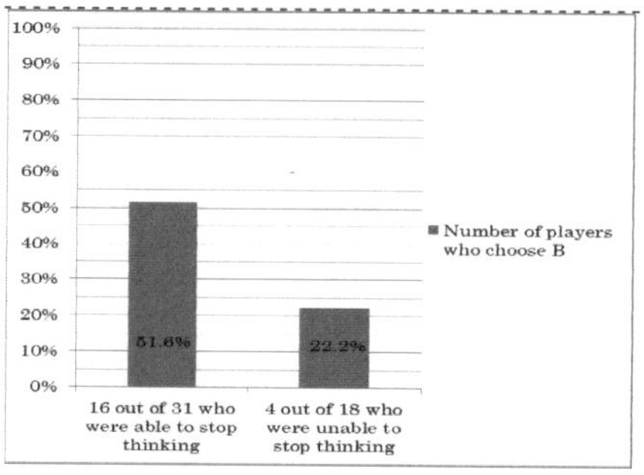

Figure 3. Expectation of subjects' own non-cooperative choice.

4. Conclusion

Why is behavior influenced by an individuals' ability to stop thinking?

Those who are unable to stop thinking are apt to devote a significant amount of thought to anticipating future events; they are circumspect and are indecisive worriers. An individual who thinks all the time might have difficulty in letting his or her worries go. In contrast, those who are able to stop thinking do not suffer from the same degree of anticipatory worry or persist in their worries. They are decisive and able to concentrate well.

We measured brain activity using a full head-type MEG device with 64 sensors. Active sites were concentrated more heavily in the central frontal region than in the occipital region while thinking was carried out, while the reverse was true while thinking was stopped.

We administered a questionnaire designed to compare subjects who were able to stop thinking with those who were not in terms of strategy choices in a dilemma game. Those without the ability to stop thinking had a tendency to opt for cooperative behavior. A dilemma game requires quick and determined decision making. It is also possible that the degree of one's ability to stop thinking is related to decisiveness.

References

[1] Benhabib J S and Nishimura K 1998 The dynamics of efficient intertemporal allocations with many agents *J. Economic Theory* **44** 301-320
[2] Pratt J W 1964 Risk aversion in the small and in the large, *Econometrica* **32** 122-136
[3] Nash J F 1951 Non-cooperative games, *Annals of Mathematics* **54** 286-295
[4] Luce R D and Raiffa H 1957 *Game and Decisions* (New York: Wiley)
[5] McNamara J M, Barta Z and Houston A I 2004 Variation in behaviour promotes cooperation in the prisoner's dilemma game, *Nature* **428** 745-748
[6] McClure S M, Laibson D I, Loewenstein G and Cohen J D 2004 Separate neural systems value immediate and delayed monetary rewards *Science* **306** 503-507
[7] Tonoike M, Nishimura K and Tobinaga Y 2006 IFMBE Proceedings **14** 2617-2621
[8] Nishimura K, Tobinaga M and Tonoike M 2008 *Progress of Theoretical Physics* **173** 332-341
[9] Nishimura K, Tonoike M and Tobinaga M 2007 Brain activity during the state in which thinking activity is stopped –as the basic research for economic psychology in *Introduction to economic psychology ed. By Koyasu M and Nishimura K* (Tokyo: Yuhikaku Publishing)
[10] Murakami S and Okada Y. 2006 Contribution 0f principal neocortical neurons magnetoencephalography and electroencephalography signals. *J Physiol.* Sep 15 **575** 925-36
[11] Cohen D 2004 Boston and history of biomagnetism *Neurology and Clinical Neurophysiology* 30:1

Acknowledgments

We would like to express our gratitude to the conference participants, especially to Professor Kazuyoshi Yoshimura and Professor Masatoshi Murase, as well as to students and colleagues who participated in this study.

This research was financially supported by the Grant-in-Aid for Scientific Research No. (S)20223001 and No. (S)19103002, JSPS, and the Global Center of Excellence (GCOE) program "Raising Market Quality — Integrated Design of 'Market Infrastructure'" sponsored by the Ministry of Education, Culture, Sports, Science and Technology.

Movement Regulation of a Sliding Actin Filament in a Reconstruction Motility Assay System

[1]I Kunita and [2]S Sakurazawa

[1]Molecular and System Life Science Unit, Riken Advanced Science Institute, 2-1 Hirosawa, Wako-shi, Saitama 351-0198, JAPAN
[2]Department of Systems Information Science, Future University Hakodate, 116-2 Kamedanakano-cho, Hakodate-shi, Hokkaido 041-8655, JAPAN

E-mail: itsuki.kunita@gmail.com

Abstract. An actomyosin system, which is the mechano-chemical systems of proteins, in a reconstruction motility assay system is one of many effective systems for discussing the formation mechanism of self-organized orders at protein levels. In this study we investigated the transformation change of sliding actin filaments using the optical microscopy technique and image analysis. As a result, the local windings propagated along an actin filament, and complex patterns of windings were formed. The three conditions of winding propagations need the following; the intermittent actions from myosins to the actin filaments, the connections between actin monomers have nonlinearity, and the actin filaments are flexible. These results suggest that the actin filaments have the regulation mechanism, which propagate the actions of myosins as windings.

1. Introduction

In biology, many kinds of complex behaviors such as periodic oscillations, or bursting, are observed at various temporal-spatial levels [1]. These biological orders are understood as one of the dissipative structures in non-equilibrium open system, which exchanged the substances, energy and information with environment. The dissipative structures depends on (i) specific nonlinear dynamics taking place in subsystems, and (ii) the way the subsystems couple. The biological system and non-biological systems, such as BZ reaction or crystal growth, have a common formative mechanism of dissipative structures. It is necessary to clarify the formation mechanism of living organisms at various temporal-spatial levels for understanding the biological machine.

Biological motions such as muscle contraction and flagellar movement, or nerve systems are the systems of living organisms, which show the complex and various behaviors. These systems have the mechano-chemical systems of proteins, and play an important function in biology. Therefore clarifying the self-organization mechanism of those systems is one of the clues for understanding living organisms.

Murase et al. investigated the self-organization phenomena on flagellar movement, which has a dynein-tubulin system as a mechano-chemical system [1]. They showed that the difference of time constants between the activated state and the inactivated state of dynein and the structural anisotropy of dynein are factors of the self-organized mechanism of movement. This result suggests that the structural characteristics or time response of proteins are important factors for complex behaviors in mechano-chemical systems of proteins.

An actomyosin system, which plays a role in muscle contractions, is also one of the mechano-chemical systems. The sliding movement of the actin filaments, which is the molecular mechanism of muscle contractions, has been studied. This sliding mechanism of the actin filament is generally explained using the Lever Arm Model, or the Loose Coupling Model. In such models, the mechanism of sliding movement was explained that the actin filaments moved by conformation changes of myosins using the energy of ATP hydrolysis. However, for details of relaxation process in conformational distortion of myosins, it is not clear whether myosins move actin filament directly, or the actin filaments move after conformation distortion of myosins is transformed into the conformation distortion of actin filaments.

In this study we discuss that the actin filaments regulate the actions of myosins by the flexion movement of the sliding actin filament. We clarify the factors of the self-organized mechanism of the flexion movement in sliding movement of actin filament. We suggest that the structural characteristics or time response of the actin filaments, which is a protein, are factors of nonlinearity in subsystems.

2. Materials and Methods

2.1. Reagents and proteins

Actin and Myosin were prepared from rabbit skeletal muscle [8]. Heavy mero-myosin (HMM) was prepared by α-chymotriptic digestion of myosin [9]. Actins were extracted from acetone powder and were purified according to the method of Spudich and Watt [10]. Fluorescent-labeled actin filaments were made and treated with equal molar rhodamin-phalloidin (Sigma Co.) in buffer A (25 mM imidazole-HCl, 25 mM KCl, 4 mM $MgCl_2$, 1 mM DTT, pH=7.4).

2.2. A reconstruction motility assay system and observations of actin filaments

We prepared a standardized reconstruction motility assay system (Figure 1.)[7]. HMMs dissolved in buffer A were perfused between a cover glass (18x18 mm) and a slide glass (24x36 mm) coated with collodion (Sigma Co.). Roughly 60 sec after the perfusion, HMMs that did not attach on the glass surface were removed with a bovine serum albumin dissolved in buffer A. Fluorescent-labeled actin filaments dissolved in buffer A were then perfused, and put on HMMs. ATP dissolved in buffer A was then perfused, and the sliding movements of actin filaments were started.

The observation of fluorescent-labeled actin filaments was performed under an optical microscope with a fluorescent illumination unit (Olympus, IX70, X100 oil objective lens, X1.5 and X2.0 relay lens). The fluorescent images were recorded on hard disk by using a personal computer with a high sensitive video camera (Hamamatsu Photonics, C2400-97V) at intervals of 1/30 sec.

2.3. Image analysis

The skeletons of actin filaments in fluorescent images were analyzed using a thinning program, written in C language with Visual Studio 2005 (Microsoft Co.), as shown in Figure 2 and Figure 3. Three consecutive data of the skeleton data of an actin filament were averaged. Here, the head-end of the filaments was defined as the end of the direction of movement, and the tail-end of the filaments was defined as the end of the backward direction of movement. The winding angle of some coordinate point P_j (x,y) along the skeleton of action filament is estimated from the following formula.

$$\theta_j = \left(\frac{180}{\pi}\right) Cos^{-1}\left(\frac{\left(P_{j-1}(x,y)-P_j(x,y)\right)\cdot\left(P_{j+1}(x,y)-P_j(x,y)\right)}{\left|\left(P_{j-1}(x,y)-P_j(x,y)\right)\right|\left|\left(P_{j+1}(x,y)-P_j(x,y)\right)\right|}\right)$$

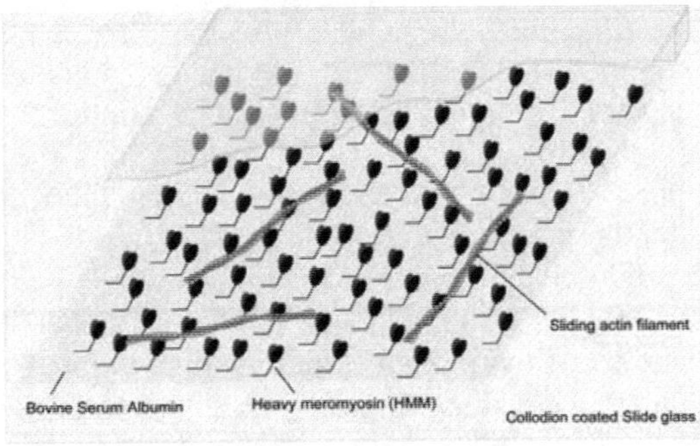

Figure 1. Illustration of a reconstruction motility assay system.

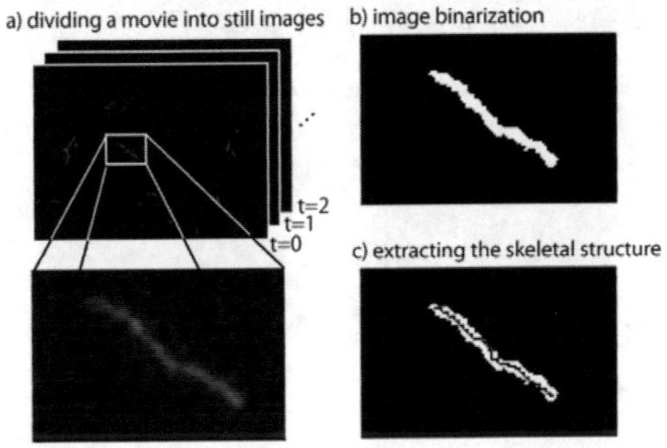

Figure 2. Processes of analysis for extracting the skeletal structure of an actin filament.

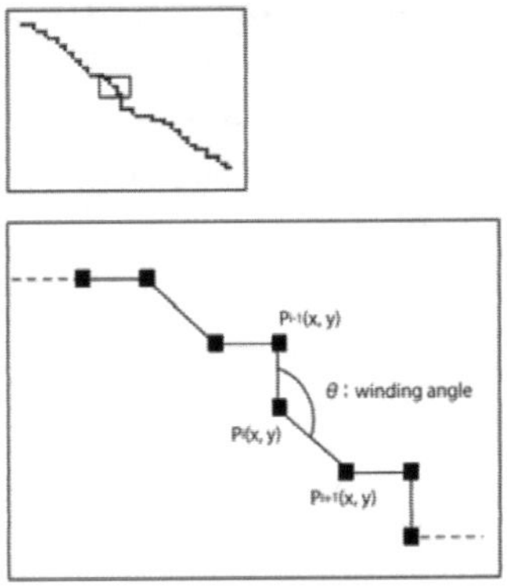

Figure 3. Definition of winding angle θ .

3. Results

3.1. Winding propagation along an actin filament

Figure 4 shows the time developments of displacement of the head-end and tail-end, and the variety of winding angles of an actin filament at every 1/30 sec. Each graph in Figure 4 shows the well known tendency that the sliding velocity increased as ATP concentration increased. The sliding velocity of actin filaments is about 1 μ m/sec at low concentration of 0.02 mM ATP, as shown in Figure 4(a), and 4∼5 μ m/sec at high concentration of 2 mM ATP, as shown in Figure 4(b). On the other hand, the time developments of each winding angle at the head-end and at the tail-end of an actin filament decreased as ATP concentration increased. Actin filaments showed slow sliding movement with frequent change in direction at low concentration of 0.02 mM ATP. Actin filaments also showed straight and smooth movement at high concentration of 2 mM ATP.

Moreover, at each concentration of ATP the change of winding angle at the tail-end was clearly large in comparison with those at the head-end. Additionally, the difference between the head-end and the tail-end tended to increase as ATP concentration increased. In other words, the number of windings along the actin filaments increased as the sliding velocity increased, but the tail-end part of the actin filament had the windings while the head-end part of the actin filament became straight.

Figure 5 shows the time development of winding angle on the skeleton of an actin filament by kymograph. The horizontal axis is time. The vertical axis is the distance from the head-end of the actin filament to a given value of the actin filament. The pixelated grayscale concentration displays the winding angle. The diagonal pattern was formed by the assembly of the high concentration pixels. It shows the second-order propagation of windings along an actin filament. Its gradient is the propagation velocity. Moving diagonally in a right upward direction displays the propagation from the head-end to the tail-end. Conversely, the right downward direction displays the propagation from the tail-end to the head-end. The area filled with light colored pixels shows the domain on a sliding actin filament in which the winding angle does not change. Generally, this domain is almost straight.

From these figures, we could see the generation of complex winding patterns along an actin filament in sliding movement. The difference of these patterns depended on ATP concentration. The fastest propagation velocity was about twice the ordinal average velocity at which the actin filaments slide at each concentration of ATP. The large windings tended to propagate in a backward direction in a high velocity sliding movement at the high concentration of ATP.

On the other hand, we found a complex phenomenon caused by changing the winding mode. For example, in Figure 5 (b), the winding propagation of the velocity of 5∼10 μ m/sec from the tail-end to the head-end reflected at the head-end point. The windings also propagated from the head-end to the tail-end at the velocity of 5∼10 μ m/sec at 0.5-2.0 sec. In Figure 5 (b), the propagation from the tail-end to the head-end switched to the propagation from the head-end to the tail-end at 1.5 sec after starting. Here, some of the winding propagations assembled after the direction of winding propagation switched their direction. After that, the whole filament become almost straight and continued its sliding movement.

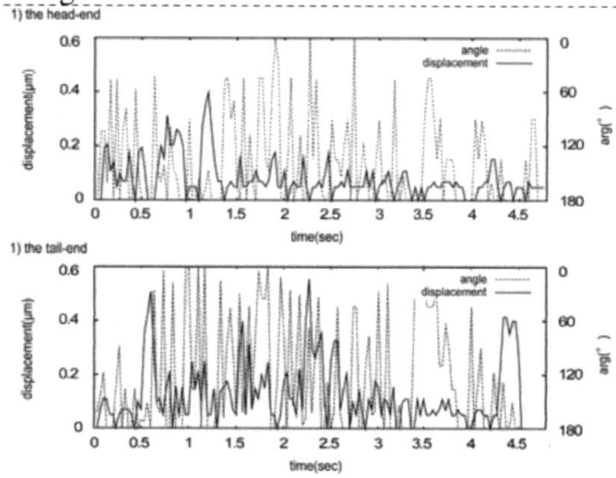

(a) [ATP] = 0.02 mM

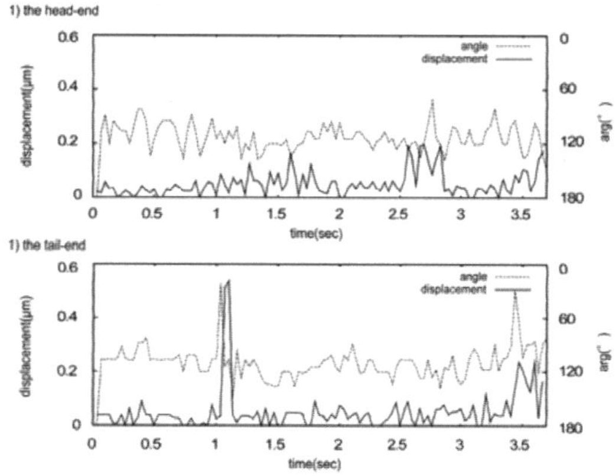

(b) [ATP] = 2.00 mM

Figure 4. Time series of displacement and change of winding angle on the head-end and the tail-end of the sliding actin filament.

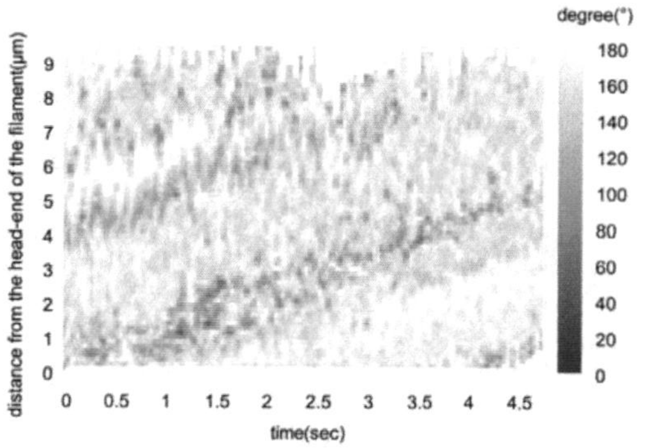

(a) [ATP] = 0.02 mM

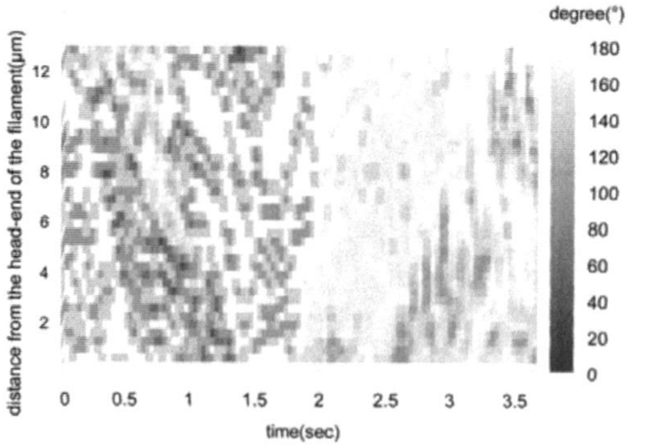

(b) [ATP] = 2.00 mM

Figure 5. Kymographs of winding propagation along an actin filament.

3.2. Winding propagation on a one-dimensional elastic body model

The windings of actin filaments occurred on both the non-sliding actin filaments and the sliding actin filaments. However, the winding propagation along an actin filament only occurred on the sliding actin filaments. This result shows that the occurrence of winding propagations needs the process in which the myosins dissociate from actin filaments in the presence of ATP. In other words, the actin filaments need the actions of activated-myosins for winding propagations. The winding propagation patterns depended on the sliding velocity of actin filaments. This result shows that the winding propagation occurs not only due to the actions of myosins, but also to the internal actions of actin filaments. Therefore, we investigated the propagation process along the local windings of actin filaments using an one-dimensional elastic body model.

$$\frac{d^2\theta_j}{dt^2} = (M_{j+1} - M_j) + n \cdot \sin\theta_j$$
$$M_j = K_l(\theta_j - \theta_{j-1}) + K_n(\theta_j - \theta_{j-1})^3$$

Here, θ is the winding angle between units, M_j is the restoring moment of the connection springs, n is the switch of external action, K_l and K_n are the spring constants, and j is a data number of one-dimensional elastic body. This one-dimensional elastic body model is dissolved by computer simulation using the Runge-Kutta method. As a result, similar local winding patterns in the sliding actin filaments were found when the model meets the following three conditions. Firstly, the external actions to the subsystem are intermittent. Secondly, the connection springs has nonlinearity. Finally, the spring constant is small.

4. Discussion

The mechanism of sliding movement of actin filament is generally explained in the models, which are based on the myosins with the ATP hydrolysis activities. In those models, the actin filaments play the role of rail, which are moved by myosins. On the other hand, Hatori and Honda et al. showed that the propagation of transversal fluctuations and acceleration along the actin filaments occur while sliding [7][11]. These results suggest that the myosin-based model is not enough to explain the movement of actin filaments. Therefore, another model is needed for the sliding movement of actin filaments.

In this study, we investigate the mechanism which regulates the actions from actin filaments to myosins. As a result, the local windings of actin filaments formed the complex patterns, and propagate along actin filaments. The propagation patterns depended on the sliding velocity or ATP concentration. The fastest propagation velocity of windings along a sliding actin filament was about twice the ordinal average velocity at which the actin filaments slide at each concentration of ATP. In addition, the windings propagate both from the head-end to the tail-end and from the tail-end to the head-end. These results show that the windings of actin filaments are not passive. In fact, these results suggest the possibility that the winding propagation of actin filaments contribute to the mechanism of sliding movement though actin filaments do not have the ATP hydrolysis activities.

We simulated the winding propagation along actin filaments by a one-dimensional elastic body model. The winding propagation along a one-dimensional elastic body was found when the model meets the three conditions stated above. Firstly, the external actions to the subsystems are intermittent. This corresponds to the intermittent actions between the actin filaments and myosins. In fact, the actin filaments associate the myosins intermittently [12]. Secondly, the connection springs has nonlinearity. This means that the actin monomers have structural anisotropy. Finally, the spring constant is small. This spring constant corresponds to the binding strength between actin monomers. The actin monomers in the actin filaments are connected with electrostatic action. As a result, the actin filaments have flexibility. These results show that the factors of winding propagation along actin filaments are the intermittent actions between the actin filaments and myosins, the nonlinearity of actin monomers, and the weak connection between actin monomers in the actin filaments.

This action between actin filaments and myosins occurs within milliseconds. On the other hand, the winding propagation of actin filaments occurs within seconds. In fact, the actions of winding propagation along actin filaments have two different time scales. This property has a resemblance to the property of other various and complex behaviours, such as flagellar movement.

These results suggest that the actin filaments have the regulation mechanism, which propagate the actions of myosins as windings to the neighboring region.

Acknowledgment

This workshop was supported in part by the Grant-in-Aid for the Global COE Program "The Next Generation of Physics, Spun from Universality and Emergence" from the Ministry of Education, Culture, Sports, Science and Technology (MEXT) of Japan.

Theoretical Physics at Kyoto University, where this work was initiated during the YITP-W-10-12 on "International and Interdisciplinary Workshop on Novel Phenomena in Integrated Complex Sciences: from Nonliving to Living Systems".

References

[1] Murase M 1992, Dynamics of Cellular Motility, John Wiley & Sons Ltd.
[2] Huxley AF and Niedergerke R 1954 *Nature* **173** 971
[3] Huxley HE and Hanson J 1954 *Nature* **173** 973
[4] Uyeda QPT, Abramson PD and Spudich JA 1996 *Proc. Natl. Acad. U.S.A.* **93** 4459
[5] Yanagida T 1990 *Adv. Biophys.* **26** 75
[6] Harada Y, Noguchi A, Kishino A, and Yanagida T 1987 *Nature* **326** 805
[7] Hatori K, Honda H, Shimada K, and Matsuno K 1998 *Biophysical Chemistry* **75** 81
[8] Perry SV 1955 *Methods in Enzymology* **2** 582
[9] Okamoto Y and Sekine T 1985 *J. Biochem.* **98** 1143
[10] Spudich JA and Watt S 1971 *The Journal of Biological Chemistry* **246** 4866
[11] Honda H, Hatori K, Igarashi Y, Shimada K, and Matsuno K 1999 *Biophysical Chemistry* **80** 139
[12] Kunita I, Sakurazawa S, Honda H 2011 BioSystems **103** 79

The effect of a 50-Hz sinusoidal magnetic field on nitric oxide (NO) production by human umbilical vein endothelial cells (HUVECs)

Hidetake Miyata[1], Ken-ichi Ishizawa[2], Masami Ishido[3], Ken Sugawara[4], Masatoshi Murase[5] and Tsuyoshi Hondou[1]

[1] Department of Physics, Graduate School of Science, Tohoku University

[2] Department of Hematology and Rheumatology, Tohoku University School of Medicine

[3] Research Center for Environmental Risk, National Institute for Environmental Studies

[4] Faculty of Liberal Arts, Tohoku Gakuin University

[5] Yukawa Institute for Theoretical Physics, Kyoto University

Corresponding author: Hidetake Miyata
Department of Physics, Graduate School of Science, Tohoku University
Sendai, Miyagi 980-8578, Japan
Tel: +81-22-795-6465; Fax: +81-22-795-6774
E-mail: miyata@bio.phys.tohoku.ac.jp

Abstract. To investigate the effect of magnetic fields on living systems, we analyzed the effect of a 50-Hz, 1-mT sinusoidal magnetic field on nitric oxide (NO) production in human umbilical vein endothelial cells (HUVECs). Statistically significant differences in NO production were identified between exposed and unexposed (sham) cells. This finding confirms previous studies showing that magnetic fields can influence NO production; however, unlike previous studies, our results demonstrate two-way deviation of NO production under the influence of a magnetic field. Thus, in some cases, NO production was higher for the exposed group than the sham group, while in other cases the reverse was true. We suggest that the magnetic field affected the homeostasis of NO levels in a complex manner. Thus, the resulting NO level was dependent on the condition of the cell in each experiment or cells in each experiment were at a different stage in the cell cycle and therefore exhibited a different response when exposed to the magnetic field.

Journal of Physics: Conference Series **344** (2012) 012006 doi:10.1088/1742-6596/344/1/012006

1. Introduction

Low-frequency (50/60 Hz) magnetic fields have been shown to interact with biological systems in a number of ways; for example, they promote healing of fractured bones and skin wounds [1]. Another study showed that a magnetic field causes single- and double-strand breaks in the DNA of brain cells of mice [2]. A magnetic field also caused frog arteriole vasodilation, a process which is regulated by nitric oxide (NO) produced by the endothelial cells that form the walls of blood vessels [3].

NO is synthesized from the amino acid L-arginine by an enzyme called nitric oxide synthase (NOS). NO is a key molecule that plays various roles in biological situations such as the immune response of macrophages, signal transduction in neurons and the regulation of blood pressure [4]. In human umbilical vein endothelial cells (HUVECs), an isoform of NOS (endothelial NOS; eNOS) is constitutively expressed and synthesizes NO (figure 1). The NO produced by HUVECs inhibits the contractility of the cell, thereby regulating blood pressure.

Several previous studies imply an intimate but apparently complex connection between magnetic fields and NO production. For example, in mouse macrophage cell lines, application of a pulsed magnetic field (10–250 Hz, 4.4 mT max) for 20 min elevated the level of NO production [5]. Similarly, in HUVECs, exposure for 20 min to an oscillating magnetic field (20 Hz, 20 mT max; generated by a rotating permanent magnet) promoted NO production [6]. In epidermal keratinocytes, application of a sinusoidal magnetic field (50 Hz, 1 mT r.m.s.) for 3 h increased NOS and NO production [7]. However, in human monocytes, overnight exposure to a sinusoidal magnetic field (50 Hz, 1 mT r.m.s.) decreased the level of NOS expression [8]. The reasons for these discrepant results remain unclear. The importance of NO in many facets of biological activities and the existence of many accessible methods for the measurement of NO prompted us to study the effect of a magnetic field on NO production by HUVECs. In the present study, we measured the level of NO produced by HUVECs using a fluorescence method and evaluated the effect of a 50-Hz sinusoidal magnetic field.

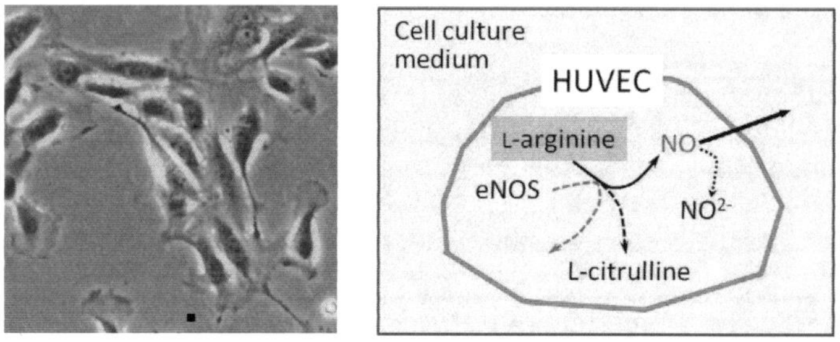

Figure 1. HUVECs and simplified scheme of nitric oxide metabolism.
A, Color photomicrograph of HUVECs. Green light (wavelength = 546 nm) was used to acquire the image. Bar, 100 μm. B, Nitric oxide (NO) is synthesized from L-arginine in HUVECs by the endothelial nitric oxide synthase (eNOS) enzyme. This enzyme acts as a catalyst and is not altered after the reaction. Some NO is converted into other chemical species such as NO_2^- in the cell cytoplasm and the remaining NO diffuses out of the cell into the cell culture medium. L-Citrulline is also a product of eNOS action.

2. Materials and methods

2.1. Cells and reagents

Journal of Physics: Conference Series **344** (2012) 012006 doi:10.1088/1742-6596/344/1/012006

HUVECs, culture medium and additives were purchased from Lonza (Cat. No. CC-2517A; Walkersville, MD, USA). Cells from the same lot were used throughout the study. For experiments, 5 × 10^5 cells were plated in a 6-cm-diameter plastic culture dish (Greiner, Tokyo, Japan) and were cultured in a CO_2 incubator (Model 5410, Precision Scientific, Chicago, IL, USA) at 37 °C (controlled to within ±0.1 °C) in 5% CO_2 for 3 or 4 days prior to the experiments.

Diaminofluorescein (DAF-2) that was used to measure the level of NO was purchased from Cayman Chemicals (Ann Arbor, MI, USA) or Sekisui Medical (Tokyo, Japan). L-arginine and other chemicals (analytical grade) were obtained from Wako (Osaka, Japan) and were used without further purification.

2.2. Apparatus and measurements

Four sets of Helmholtz coils (inner diameter, 115 mm; thickness, 40 mm; Model U21901, 3B Scientific GmbH, Hamburg, Germany) were used for each experiment (figure 2A and B). Two sets were placed side-by-side in incubator 1 and the other two sets were placed in incubator 2. As shown in figure 2C, sinusoidal current generated by a function generator (Model WF1943A, NF Co., Yokohama, Japan) was fed to each coil through an electric amplifier (Model PMA390AE, DENON, Kawasaki, Japan). A Tesla meter (Model 421, Lake Shore Cryotronics, Inc., Westerville, OH, USA) was used to measure the magnetic flux density at the center of the coil pair.

Thermometers consisting of a thermistor probe and a data logger (TR-77Ui, T&D Co., Matsumoto, Japan) were used to monitor and record transient temperature changes in each coil. The thermistor was placed near the bottom of the culture dish (figure 3A). Data of transient temperature changes such as those shown in figure 3B were used to calculate the temperature difference between the exposed and the sham coils (see Data analysis).

The mechanical vibration of the coil base was measured using a motion sensor (sensor 2304A connected to the recorder 1332B; Showa Sokki, Tokyo, Japan; figure 4). The sensor was attached with a double-sided adhesive tape to the front edge of the coil base (figure 4, cross symbol) and the motion of the support was recorded for 4.4 sec at 50 kHz.

NO production was quantified using the fluorescent NO probe DAF-2, whose fluorescence intensity increases about 100-fold when it reacts with NO [9, 10]. The fluorescence intensity of DAF-2 was measured using a spectrofluorophotometer (Model FP-750, JASCO, Tokyo, Japan). The excitation and emission wavelength were 495 and 515 nm, respectively, with a bandwidth of 5 nm in both channels.

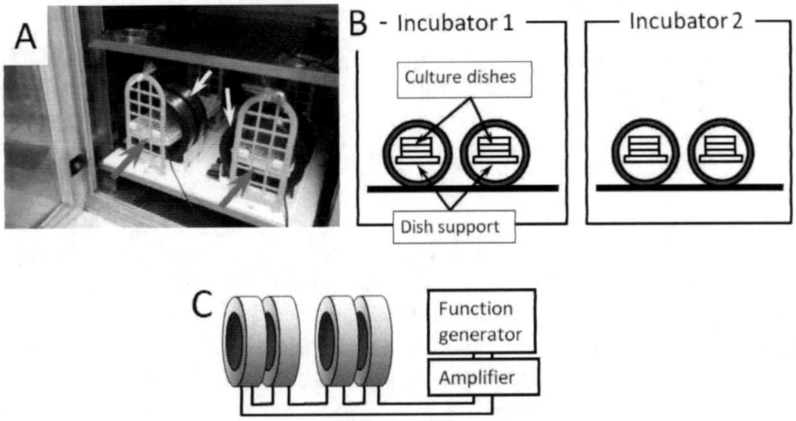

Figure 2. Apparatus used for the experiments.

A, A photograph of two sets of Helmholtz coils (white arrows) placed side by side in a CO_2 incubator. The gray arrows indicate supports for cell culture dishes. B, A schematic representation of the arrangement of the Helmholtz coils and cell culture dishes in two CO_2 incubators. C, Two sets of Helmholtz coils are connected in series to an amplifier and a function generator.

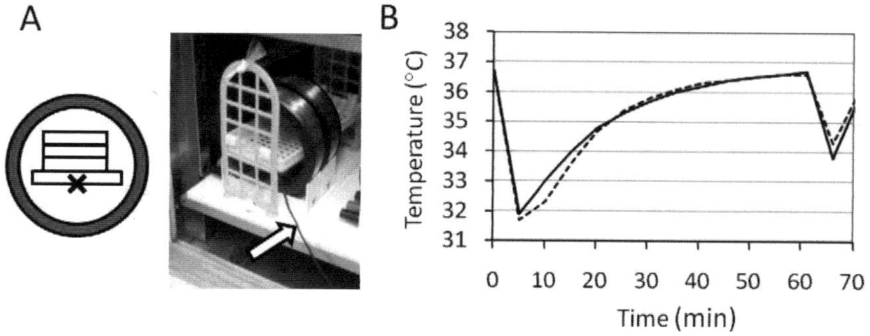

Figure 3. Measurement of transient temperature changes.
A, Measurement of temperature in the Helmholtz coil: the probe was attached to the lower side of the dish support, as indicated by a cross in the left panel. The white arrow in the right panel indicates the cable connecting the thermistor sensor to the data logger.
B, An example of a temperature record. Solid and dotted lines represent the record obtained in exposed and sham coils, respectively. The door of the incubator was opened at time zero. The average temperature was calculated during the period of temperature restoration (see Data Analysis).

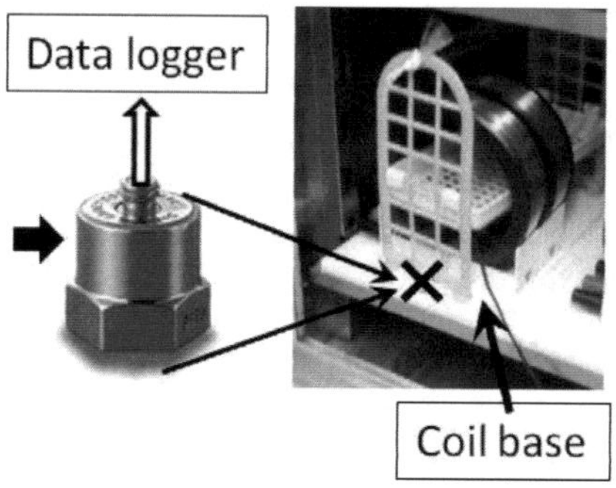

Figure 4. Measurement of mechanical vibration.
A motion sensor (black arrow) was attached to the front of the base of the Helmholtz coil (cross) to detect vibration arising from the active coil.

2.3. Exposure

On the day of each experiment, the cells were washed twice in Krebs-Ringer phosphate buffer (KRP buffer). Then, KRP buffer containing 1 mM L-arginine and 1 mM DAF-2 was added to each culture dish. For exposure to the magnetic field, two or three culture dishes were placed on a dish support in each coil (figure 2A and B) in incubator 1 and an electric current was applied to the coil (figure 1C); dishes containing cells for the unexposed (sham) group were placed in a Helmholtz coil in incubator 2 without applying current. Cells were then left in the incubators for 1 h, during which time the temperature in the coil was monitored using the thermistor probe. The temperature difference between the two incubators in the steady state was maintained within 0.2 °C. After application of the magnetic field, the culture medium in each dish was collected in separate test tubes and fluorescence intensities were measured using a spectrofluorophotometer.

2.4. Data analysis

The florescence intensities of individual exposed and sham samples were ensemble-averaged and are respectively designated as $<I_{exp}>$ and $<I_{sham}>$. Likewise, the ensemble-averaged florescence intensities of individual sham 1 and sham 2 samples (incubated in the incubator 1 and 2, respectively) were designated as $<I_{sham1}>$ and $<I_{sham2}>$. Fluorescence intensity was statistically analyzed using Welch's test. The p values representing statistical significance were calculated for each set of ensemble-averaged values, and $p < 0.05$ was judged as statistically significant.

In order to calculate the difference in temperature between exposed and sham coils, T_{Exp} and T_{Sham}, or that between two sham coils, T_{Sham1} and T_{Sham2}, we calculated the time-average of the temperature values, T^*_{Exp}, T^*_{Sham}, T^*_{Sham1} and T^*_{Sham2}, and the differences, $T^*_{Exp} - T^*_{Sham}$ and $T^*_{Sham2} - T^*_{Sham1}$ were taken.

The data of the displacement record of the coil base was Fourier-transformed and a power spectrum was calculated using the program supplied by the manufacturer of the motion sensor.

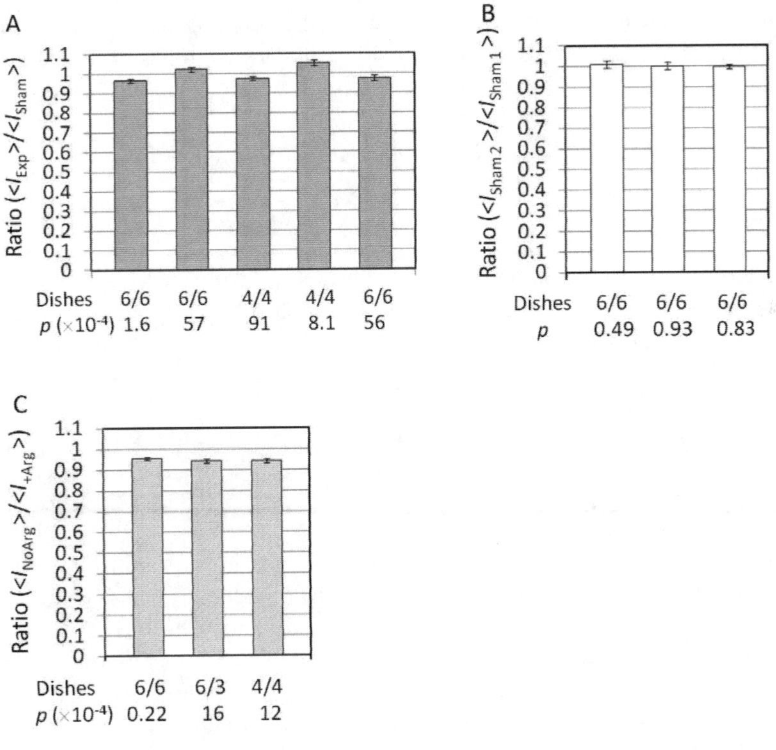

Figure 5. Effect of a magnetic field on nitric oxide production in HUVECs.
A, Summary of $<I_{Exp}>/<I_{Sham}>$ values obtained in five independent experiments. Fluorescence intensity of the cell culture medium was assayed. The error bar was plotted using standard deviation. "Dishes": number of dishes assayed per experiment; p: statistical significance of the difference between $<I_{Exp}>$ and $<I_{Sham}>$ values for each experiment. B, The result of three independent, all-sham experiments; ratios of $<I_{Sham2}>$, obtained from incubator 2, to those of $<I_{Sham1}>$ from incubator 1 are shown. The symbols are as defined in A, except that in the row labeled "Dishes" the numerals to the left and the right of the thrash indicate the number of samples for Sham 1 and for Sham 2, respectively.

C, Ratio of fluorescence intensity in the absence of L-arginine ($<I_{NoArg}>$) to that in the presence of L-arginine ($<I_{+Arg}>$), obtained from three independent experiments. In the row labeled "Dishes" the numerals to the left and the right of the thrash indicate the number of samples with and without L-arginine, respectively.

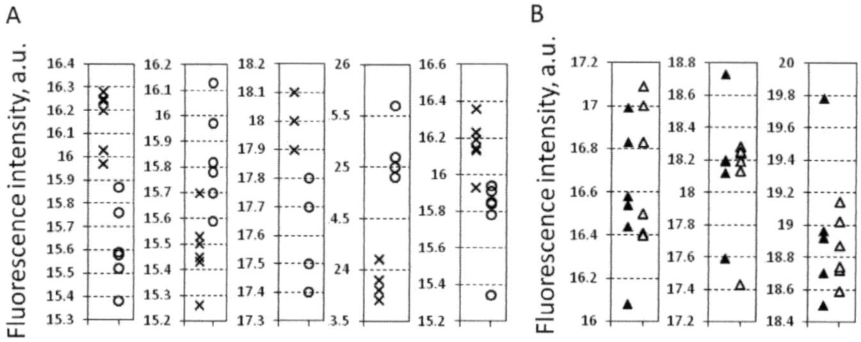

Figure 6. Distribution of individual I_{Exp} and I_{Sham} values.
A, Distribution of fluorescence intensities of exposed and sham samples. Open circles, exposed; crosses, sham. B, Distribution of fluorescence intensities of samples measured in incubator 1 (closed triangles) and in incubator 2 (open triangles), both of which were measured without exposure to a magnetic field.

3. Results and Discussion

Cellular NO production was assessed using a fluorescent method from cells incubated with L-arginine and exposed, or not exposed, to a magnetic field. Figure 5A shows $<I_{exp}>/<I_{sham}>$ values with standard deviation, which were obtained from five independent experiments. Individual differences from unity were small, but they were statistically significant, as indicated by the p values obtained in individual experiments (p values are in the bottom row in figure 5A). These differences were not due to the use of two separate incubators for exposure and sham conditions, because when all coils in the two incubators were operated without current (all sham conditions), the resultant $<I_{Sham1}>/<I_{Sham2}>$ values, i.e., the ratio of fluorescence intensities obtained from incubators 1 and 2, exhibited no significant difference from unity, as shown in figure 5B.

We also carried out an experiment to examine the effect of omission of L-arginine, the precursor of NO, under all-sham conditions. The fluorescence intensity observed in the presence of L-arginine $<I_{+Arg}>$ was about 5% higher than that that observed in the absence of L-arginine ($<I_{NoArg}>$; figure 5C). The level of fluorescence intensity in the absence of L-arginine should represent the basal level of DAF-2 fluorescence. Therefore, the 3-5% difference between $<I_{Exp}>$ and $<I_{Sham}>$ values obtained in the presence of L-arginine was rather large in comparison to the 5% difference between $<I_{+Arg}>$ and $<I_{NoArg}>$ values.

The above difference between $<I_{Exp}>$ and $<I_{Sham}>$ values is a reflection of the characteristic distribution of individual I_{Exp} and I_{Sham} values. Namely, individual I_{Exp} and I_{Sham} values in each set of experiment exhibit few overlaps (figure 6A). On the other hand, the data points representing I_{Sham1} and I_{Sham2} exhibited considerable overlap (figure 6B).

We next evaluated the effect of a potential difference in temperature in the exposed and sham coils. Temperature in the coils quickly dropped when the door of the incubator was opened in order to place the culture dishes in the coil, and the temperature gradually approached the original level after the door was closed, as shown in figure 3B. It was possible that temporal changes in exposed and sham coils might be different and could result in differences in NO production by the eNOS as in general, enzyme activity is sensitive to the ambient temperature. Therefore, as a measure of the temperature difference, we calculated the difference in time-averaged temperature values, $T^*_{Exp} - T^*_{Sham}$, as described in Data Analysis. We plotted $T^*_{Exp} - T^*_{Sham}$ against $<I_{Exp}> - <I_{Sham}>$, or $T^*_{Sham2} - T^*_{Sham1}$ against $<I_{Sham2}> - <I_{Sham1}>$ (figure 7). It is apparent that the plots are scattered. Therefore, the differences in temperature values of incubator 1 and 2 were unlikely to be the reason for the difference in fluorescence intensities measured in the two incubators.

We then examined if the vibration of the active Helmholtz coil affected HUVEC function, because HUVECs are known to respond to shear stress of the blood stream by altering NO production [11]. The vibration of the coil would be transmitted to the culture dishes through the dish support placed on the base of the coil. We therefore measured the vibration of the coil base using the motion sensor. As shown in figure 8, the power spectrum of the active coil and that of the inert coil were superimposable; the large peak around 23 Hz probably arose from the rotation of a fan for circulation of air in the incubator, because this peak appeared under both active and inert conditions. We conclude that mechanical vibration did not cause the difference between $<I_{Exp}>$ and $<I_{Sham}>$ values.

On the basis of the results shown in figure 5, and upon examination of possible confounding factors, i.e., temperature difference and mechanical vibration, we conclude that the magnetic field did affect the level of NO production in HUVECs. Our result is consistent with previous studies that used HUVECs and other types of cells.

As shown in figure 6A, the plots of I_{Exp} and those of I_{Sham} were almost completely separated. It could be argued that this separation is a chance phenomenon. However, we suggest that this is unlikely based on the following consideration: the chance that 4 plots from exposed samples and 4 plots from sham samples will be separated is 0.023 ($= 2 \times 4! \times 4!/8!$). Hence, the probability of the occurrence of such a phenomenon in several independent experiments is very small. We consider that the separation did not occur by chance and that it was a result of interaction between the magnetic field and HUVECs.

Our result shows that $<I_{sham}>$ was larger than $<I_{exp}>$ in some experiments, whereas it was smaller in other experiments (figure 5A). In previous studies, application of a magnetic field was found to increase the level of endothelial NOS in human keratinocytes [7], whereas in another study it reduced expression of NOS in human monocytes [8]. These disparate results are often attributed to differences in the experimental conditions (e.g., cell type or the pattern of the applied magnetic field). However, in our case the disparate results were obtained under the same experimental conditions. We consider that the cellular response to the magnetic field may be dependent on the stage of the cell cycle (e.g., cell growth and division, DNA replication), as was suggested in previous studies [12]. The level of NO is determined by the balance between production and conversion into other chemical species. In general, this homeostasis involves negative feedback, which is essential in biological systems [13]. Negative feedback results from signal transduction pathways that usually form interconnected networks and “cross talk” with each other [14]. Complex signal transduction pathways can show quite divergent responses, including directionally opposite responses, even when an identical stimulus is applied [15]. In our case, the magnetic field may alter NO production rate depending on the stage of the cell cycle. However, the magnetic field may also alter the conversion rate. This hypothesis would explain our results because cells are likely to be distributed in multiple stages in the cell cycle and this distribution is probably not identical in individual experiments. Alternatively, the magnetic field might have affected the feedback mechanism, but this effect may be different at different stages and, in some stages, the feedback may “over-compensate” for the increased NO production. If so, the different distribution of stages again determine the final NO level. At present, we cannot evaluate these possibilities. Our future experiments aim to evaluate these hypotheses.

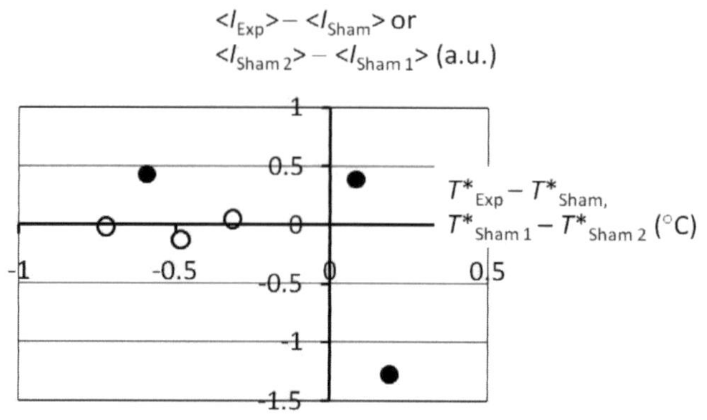

Figure 7. Analysis of temperature differences in exposed and sham coils.
Closed circles: plots of difference in the ensemble-averaged fluorescence intensity of exposed and sham (<I_{Sham}> – <I_{Exp}>) against the time-averaged difference in temperature of exposed and sham ($T^*_{Sham} - T^*_{Exp}$). Open circles: similar plots, but the differences in fluorescence intensities (<I_{Sham1}> – <I_{Sham2}>) and the temperature differences, $T^*_{Sham2} - T^*_{Sham1}$, were calculated between sham samples in incubators 1 and 2.

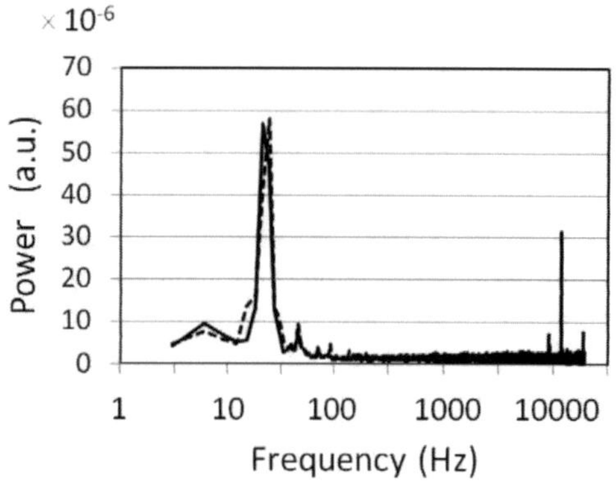

Figure 8. Power spectra of mechanical vibration.
Power spectra were calculated based on the data collected using the motion sensor; the output was in volts. Dotted line, measured spectrum of the inert coil; solid line, spectrum of the active coil.

Acknowledgements:

We thank Professor Shun Murabayashi from Hokkaido University, Dr. Mikio Miyata from Kitasato Institute Hospital and Mitsugu Sakabe from Tokai University for their helpful comments and discussions.

M. I., T. H. and M. M. would like to acknowledge the Yukawa Institute for Theoretical Physics for valuable discussions at the three successive international symposiums entitled "What is Life?" (NY2007) in 2007, "What is Creativity?" (YITP-W-08-11) in 2008, and "What is Evolution?" (YITP-W-09-14) in 2009.

All the authors further thank the Yukawa Institute for Theoretical Physics at Kyoto University. Discussions during the YITP workshop YITP-W-10-12 on "International and

Interdisciplinary Workshop on Novel Phenomena in Integrated Complex Sciences: from Non-living to Living Systems" were useful to complete this work.

This work was supported by a grant-in-aid from the Ministry of Education, Culture, Sports, Science and Technology, Japan to T. H. (No. 21200025) and H. M. (No. 20200024).

References:

[1] Diniz P, Soejima K and Ito G 2002 Nitric oxide mediates the effects of pulsed electromagnetic field stimulation on the osteoblast proliferation and differentiation *Nitric Oxide* **7** 18-23

[2] Lai, H and Singh, NP 2004 Magnetic-field-induced DNA strand breakes in brain cells of the rat. *Env. Health Perspect.* **112** 687-94

[3] Miura M, Takayama K and Okada J 1993 Increase in nitric oxide and cyclic GMP of rat cerebellum by radio frequency burst-type electromagnetic field radiation. *J. Physiol.* **461** 515-24

[4] Moncada S, Palmer RM and Higgs EA 1991 Nitric oxide, physiology, pathophysiology, and pharmacology *Pharmacol. Rev.* **43** 109-42

[5] Murabayashi S 1998 Physical approach for immunomodulations. *Tissue Culture Eng.* (in Japanese) **24** 427-31

[6] Sakamoto N, Ohashi T and Sato, M 2002 Effect of magnetic field on nitric oxide synthesis of cultured endothelial cells Int. *J. Appl. Electromag. Mech.* **14** 317-22

[7] Patruno A, Amerio P, Pesce M, Vianale G, Di Luizio S, Tulli A, Franceshelli S, Grilli A, Muraro R and Reale M 2009 Extremely low frequency electromagnetic fields modulate expression of inducible nitric oxide synthase, endothelial nitric oxide synthase and cyclooxygenase-2 in the human keratinocyte cell line HaCat: potential therapeutic effects in wound healing Brit. J. Dermatol. 162 258-66

[8] Real M, De Lutiis MA, Patruno A, Speranza L, Felaco M, Grilli A, Macri MA, Comani S, Conti P and Di Luzio S 2006 Modulation of MCP-1 and iNOS by 50-Hz sinusoidal electromagnetic field *Nitric Oxide* **15** 50-7

[9] Kojima H, Nakatsubo N, Kikuchi K, Kawahara S, Kirino y, Nagoshi H, Hirata Y and Nagano T 1998 Detection and imaging of nitric oxide with novel fluorescent indicators: diaminofluoresceins *Anal. Chem.* **70** 2448-53

[10] Wardman P 2007 Fluorescent and luminescent probes for measurement of oxidative and nitrosative species in cells and tissues: Progress, pitfalls, and prospects *Free Radical Biol. Med.* **43** 995-1022

[11] Noris M, Morigi M, Donadelli R, Aiello S, Foppolo M, Todeschini M, Orisio S, Remuzzi G and Remuzzi A 1995 Nitric oxide synthesis by cultured endothelial cells is modulated by flow conditions *Circ. Res.* **76** 536-43

[12] Findl E 1987 Membrane transduction of low energy level fields and the Ca^{++} hypothesis *Mechanistic Approaches to Interactions of Electric and Electromagnetic Fields with Living Systems* (New York) 15-38

[13] Luben RA 1995 Membrane signal-transduction mechanisms and biological effects of low-energy electromagnetic fields. Martin Blank (ed.): Electromagnetic Fields *Advances in Chemistry* Ser. 250. (Washington DC) 437-50

[14] Hung S 2000 Regulation of cellular states in mammalian cells from a genomewide view *Gene Regulation and Metabolism*. (Cambridge, MA) 181-220

[15] Murase M 2008 Environmental pollution and health: an interdisciplinary study of the bioeffects of electromagnetic fields *SANSAI*, **No.3** 1-35

IIW-NPICS 2010 IOP Publishing
Journal of Physics: Conference Series **344** (2012) 012007 doi:10.1088/1742-6596/344/1/012007

Transcriptional effects of 50 Hz magnetic fields at 1.2 μT and 100 μT on human breast cancer MCF-7 cells

Masami Ishido[1], Hidetake Miyata[2], Ken-ich Ishizawa[3], Masatoshi Murase[4], and Tsuyoshi Hondou[2]

[1]Environmental Risk Res Program, National Institute for Environmental Studies, Tsukuba 305-8506, Japan

[2]Department of Physics, Graduate School of Science, Tohoku University, Sendai 980-8578, Japan

[3]Department of Haematology and Rheumatology, Tohoku University, Graduate School of Medicine, Sendai 980-8578, Japan

[4] Yukawa Institute for Theoretical Physics, Kyoto University, Kyoto 606-8502, Japan

E-mail: ishidou@nies.go.jp

Abstract. The International Agency for Research on Cancer (IARC) classified power frequency magnetic fields as a possible human carcinogen. Alteration in transcription programs is a fundamental feature of cancer. Here, using DNA array technology, we examined the transcriptional effects of 50 Hz magnetic fields on human breast cancer MCF-7 cells. It was found that expression of several oncogenes was significantly altered by magnetic-field exposure and that gene expression profilings were similar in MCF-7 cells exposed to magnetic fields at 1.2 μT and 100 μT for 1 week.

1. Introduction

Epidemiological data reported by Wertheimer and Leeper stimulated the research area of biological effects of magnetic fields [1]. They first identified the association between child leukaemia and residential magnetic fields of extremely low frequency. They also reported a nearly threefold increase of breast cancer risk among women younger than 55 who lived near power lines, suggesting that magnetic fields exposure had accelerated development and growth of breast cancer. Subsequent epidemiological studies showed the association between child leukemia and residential magnetic fields [2-6]. The International Agency for Research on Cancer classified power frequency magnetic fields as a possible human carcinogen [7].

Stevens hypothesized that magnetic fields affect pineal gland melatonin secretion *in vivo*, which, in turn, influence mammary (breast) carcinogenesis [8]. Since then, a number of experimental studies have been conducted in order to test this hypothesis.

Melatonin from the pineal gland is released into the blood stream and cerebrospinal fluid. Melatonin was initially found to function as a mediator of circannual reproductive rhythms as well as of circadian cycles [9]. Subsequently, melatonin was shown to have significantly broader actions including oncostatic effects, immune enhancement, and anti-inflammatory functions [10, 11]. Melatonin appears to exert the effects through pharmacologically specific, high affinity receptors. Receptor affinity is sensitive to guanine nucleotides and activation of these receptors results in inhibition of adenylyl cyclase via a pertussis toxin-sensitive mechanism (Fig. 1).

Liburdy originally demonstrated that 60 Hz magnetic fields at 1.2 μT inhibit the antiproliferative effect of melatonin on the MCF-7 cells [12]. In addition, other laboratory independently reported the results that are consistent with this finding concerning the effect of magnetic fields on melatonin [13], supporting the hypothesis. We extended their study by showing the molecular mechanism of the biological effects of magnetic fields. Namely, exposure to magnetic fields causes the uncoupling of the melatonin signal transduction pathway [14, 15]. It is likely that magnetic fields certainly exert its effects on protein levels.

Alteration in transcription programs is a fundamental feature of cancer. Therefore, in this study we employed human DNA arrays of Atlas Glass Human 1.0 Microarray (Clontech Corp), containing 1,081 probe sets for transcriptional analyses of bioeffects of magnetic fields.

2. Materials and Methods

2.1. **Cell culture and Magnetic fields exposure.** MCF-7 cells were kindly provided by Dr. Liburdy (UCLA, Berkley) and grown in Dulbecco's modified Eagle's Medium (Invitrogen, Carlsbad, CA) supplemented with 10% FBS (Gibco BRL), penicillin (100 U/ml), and streptomycin (100 μg/ml) in a humidified atmosphere of 95% air: 5% CO_2 at 37°C, as described previously [14-16]. The cells were subcultured (1:4) 1 to 2 times per week. In order to expose the cells to magnetic fields, a 50 Hz sinusoidal magnetic field was generated in a chamber with four Merritt-coil devices. The chamber had four ventilation holes (2.54 cm in diameter) on the top and bottom. A temperature probe was placed inside the chamber to monitor temperature continuously. The anti-parallel mode of operation generated opposing magnetic fields that cancelled and resulted in a true sham exposure. When a current was applied to the parallel configuration, a magnetic field was established. Two identical exposure systems were employed in this study. Each coil system was driven by identical signal generators obtained from NF Electronic Instruments Corp. (Yokohama, Japan). Cell viability was determined by means of the crystal violet staining method.

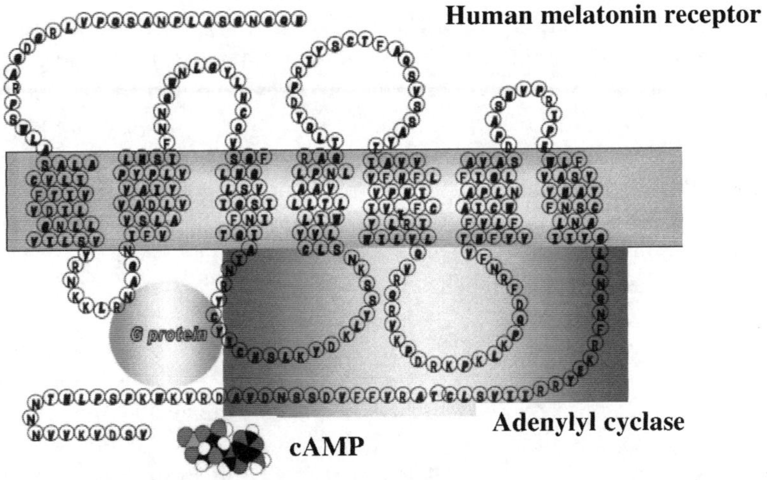

Fig. 1. The schematic drawing of melatonin signalling pathway. Melatonin acts through its receptor, G protein, and adenylyl cyclase to elicit the physiological responses such as circadian rhythms, seasonal reproduction, and oncostatic effects. This pathway was specifically disrupted by exposure to magnetic field. The primary structure of human melatonin receptor is represented with deduced amino acids (a single letter).

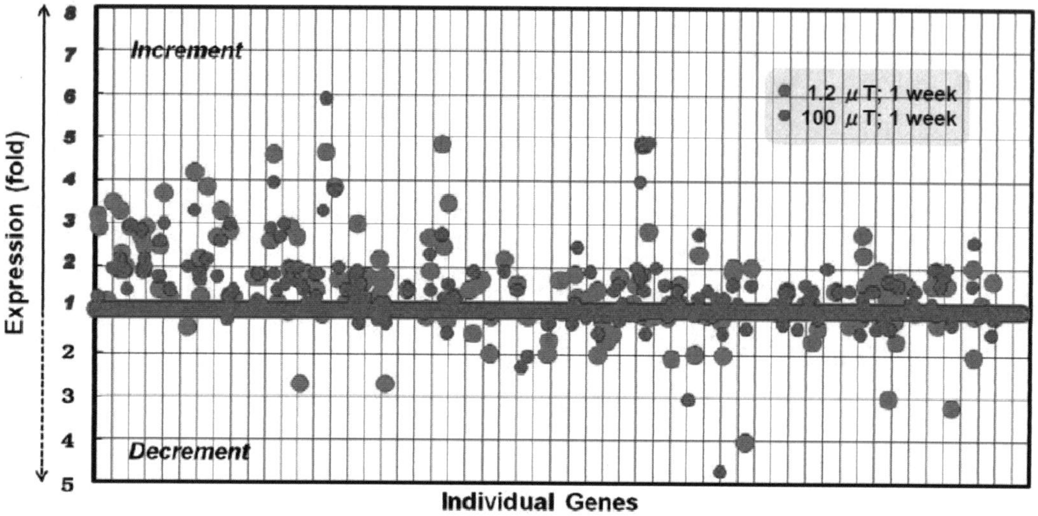

Fig. 2. Gene expression profiling of exposure to 50 Hz magnetic fields at 1.2 μT or 100 μT for 1 week. Total RNA was isolated from control or MCF-cells exposed to magnetic fields. Fluorescence-labeled cDNA probes were synthesized and hybridized with Atlas Glass Human 1.0 Microarray (Clontech). Gene expression fold was calculated as the ratio of the control cells. Expression 1 fold was represented as 'unchanged'. Genes in x-axis were functionally categorized. Note that for individual genes only estimated values are given.

2.2. **DNA array analyses.** Total cellular RNA was isolated from control and magnetic fields-treated cells (1.2 μT or 100 μT; one week) by single-step guanidinium thiocyanate-phenol-chloroform extraction, as described previously [17]. The quality and quantity of RNA was measured by spectrophotometry and electrophresis on denaturing agarose gel. The first strand DNA was synthesized using Moloney murine leukemia virus reverse transcriptase (Invitrogen) with random heximers as primers. Fluorescent-labeled cDNA probes were synthesized from 10 μg total RNA in the presence of dATP, dCTP, dTTP, dGTP, aminoally-dUTP, Cy3 and Cy5 mono-reactive dyes (Amersham Pharmacia Biotech, NJ), and PowerScript reverse transcriptase (Clontech). The labeled cDNA was purified using Atlas NucleoSpin Extraction Spin Columns (Clontech). Atlas Glass Human 1.0 Microarray (Clontech) were prehybridized in GlassHyb hybridization buffer, containing the sheared salmon testes DNA for 30 min at 50 °C in a hybridization chamber (Clontech). The heat-denatured probe was hybridized overnight at 50 °C. The arrays were washed with 2 x SSC (1 x SSC = 150 mM NaCl/ 15 mM sodium citrate, pH 7.0) twice at room temperature for 5 min, twice with 2 x SSC plus 1% SDS at room temperature for 30 min, and twice with 0.1 x SSC plus 0.5% SDS at room temperature for 30 min. The arrays were then scanned. Image acquisition and quantification were performed using an AtlasImage (Clontech).

3. Results and Discussion

The DNA array was performed using Atlas Glass Human 1.0 Microarray. There were 1,081 probe sets. The probe sets were categorized into oncogenes and tumor suppressor genes, DNA damage repair-related genes, cell cycle-related genes, cell signalling-related genes, ligand genes, and channel and transporter genes.

Gene expression levels were calculated as the ratio of expression level in the exposed cells to that in the control cells. Only estimated levels of gene expression are shown in Fig. 2. Table 1 summarizes the changes of levels in gene expression of molecules that are oncogenes or tumor suppressor genes. Genes listed were functionally categorized according to the information provided by BD Biosciences Clontech. The levels of gene expression of 95% genes were unchanged. The expression levels of 39 genes were increased by more than 2-fold, whereas that of 12 genes were decreased by 2-fold or less as a result of magnetic fields exposure (1.2 μT for 1 week). The similar changes in patterns and the extent of gene expression were obtained by exposure to 100 μT magnetic fields for 1 week. The highest change was about 6-fold increase.

<Table 1> Gene expression of typical oncogenes altered by magnetic fields exposure in MCF-7 cells, using Atlas Glass Human 1.0 Microarray

Oncogene	1.2 μT	100 μT
	-fold	
1. B-MYB proto-oncogene	1.8	2.1
2. c-jun proto-oncogene	2.7	4.1
3. Myc proto-oncogene	1.9	3.3
4. c-cbl proto-oncogene	2.5	2.7
5. Ets proto-oncogene	1.8	2.1
6. B-raf proto-oncogene	2.0	1.8
7. c-kit proto-oncogene	4.4	6.3
8. Met proto-oncogene	6.4	5.1
9. ETR 101	4.7	4.7

It has been shown that among these proto-oncogenes, the expression of Ets oncogenes was induced by magnetic fields in several kinds of cultured cells [18]. Ets is a family of transcription factors that play important roles in cell development, cell differentiation, apoptosis, and tissue remodelling [19, 20]. The aberrant expression of Ets genes leads to malignant transformation and tumor progression. The level of Ets oncogene has been associated with the grade of malignancy and prognosis in several types of tumors including breast cancer [19, 21].

The mechanism through which magnetic fields modulate gene expression is largely unknown because in single photon interaction picture, it has been considered difficult to explain the mechanism, due to its low energy. However, it has been proposed that weak magnetic fields can cause charge movement, resulting in conformation change of the macromolecules due to redistribution of charges [22]. Thus, it might be possible that magnetic fields may alter biological processes such as gene expression through their effects on charge distribution.

Since Atlas Glass Human 1.0 Microarray used in this study has limited in number of probe sets (only 1,081), it is necessary to establish the transcriptome of magnetic fields-exposed MCF-7 cells, using whole human genome DNA array. This is under way in our laboratory.

4. Conclusion

Based on 'Melatonin Hypothesis'[7], we previously demonstrated that magnetic fields exerted its effects on the protein levels in cultured MCF-7 cell, seen as a disruption of protein-protein interactions in a melatonin-signaling pathway. Furthermore, in this study we demonstrated the effects of magnetic fields on transcriptional levels by showing that the magnetic field increased expression of some oncogenes and was decreased some of others. The patterns of the change were similar at 1.2 μT and 100 μT, eliminating the possibility of artifacts during exposure to magnetic fields. The pathological consequences of these alterations remain to be determined.

References

[1] Wertheimer N and Leeper E 1979 *Am. J. Epidemiol.* **109** 273
[2] Ahlbom A, Day N, Feychting M, Roman E, Skinner J, Dockerty J, Linet M, McBride M, Michaelis J, Olsen J H, Tynes T, and Verkasalo P K 2000 *Br. J. Cancer* **83** 692
[3] Greenland S, Sheppard A R, Kaune W T, Poole C, and Kelsh M A 2000 *Epidemiol.* **11** 624
[4] Schuz J, Grigat J P, Brinkmann K, and Michaelis J 2001 *Int. J. Cancer* **91** 728
[5] Draper G, Vincent T, Kroll M E, and Swanson J 2005 *BMJ* **330** 1
[6] Kabuto M, Nitta H, Yamamoto S, Yamaguchi N, Akiba S, Honda Y, Hagiwara J, Isaka K, Saito T, Ojima T, Nakamura Y, Mizuno T, Ito S, Eboshida A, Yamazaki S, Sokejima S, Kurokawa Y, and Kubo O 2006 *Int. J. Cancer* **119** 643

[7] IARC 2002 *IARC Monogr. Eval. Carcinog. Risk Hum.* **80** 1
[8] Stevens R G 1987 *Am. J. Epidemiol.* **125** 556
[9] Borjigin J, Li X, and Snyder S 1999 *Ann. Rev. Pharmacol. Toxicol.* **39** 53
[10] Kerenyi N A, Pandula E and Feuer G M 1990 *Drug Metab. Drug Interact.* **8** 313
[11] Reiter R J 2010 *Pro.g Brain. Res.* **181** 127
[12] Liburdy R P 1993 *J. Pineal Res.* **14** 89
[13] Blackman C F, Benane S G, and House D E 2001 *Bioelectromagnetics* **22** 122
[14] Ishido M, Nitta H, and Kabuto M 2001 *Carcinogenesis* **22** 1043
[15] Ishido M and Kabuto M 2003 *Environ. Sci.* **10** 349
[16] Ishido M 2004 *Environ. Sci.* **11** 163
[17] Ishido M 2010 *J. Mol. Neurosci.* **42** 349
[18] Romano-Spica V, Mucci N, Ursini C L, Ianni A, and Bhat N K 2000 *Bioelectromagnetics* **21** 8
[19] Oikawa T 2004 *Cancer Sci.* **95** 626
[20] Turner D P, Findlay V J, Moussa O, and Watson D K 2007 *J. Cell Biochem.* **102** 549
[21] Span P N, Manders P, Heuvel J J, Thomas C M Bosch R R, Beex L V and Sweep C G 2002 *Oncogene* **21** 8506
[22] Blank M 2008 *Electromagn. Biol. Med.* **27** 3

Acknowledgment

The authors thank the Yukawa Institute for Theoretical Physics at Kyoto University. Discussion during YITP workshop YIT-W-10-12 on "International and Interdisciplinary Workshop on Novel Phenomena in Integrated Complex Sciences: from Non-living to Living Systems" were useful to complete this work.

IIW-NPICS 2010 IOP Publishing
Journal of Physics: Conference Series **344** (2012) 012008 doi:10.1088/1742-6596/344/1/012008

A spatiotemporal structure: common to subatomic systems, biological processes, and economic cycles

Ken Naitoh
Faculty of Science and Engineering, Waseda University,
3-4-1 Ookubo, Shinjuku, Tokyo 169-8555 Japan

E-mail: k-naito@waseda.jp

Abstract. A theoretical model derived based on a quasi-stability concept applied to momentum conservation (Naitoh, JJIAM, 2001, Artificial Life Robotics, 2008, 2010) has revealed the spatial structure of various systems. This model explains the reason why particles such as biological cells, nitrogenous bases, and liquid droplets have bimodal size ratios of about 2:3 and 1:1. This paper shows that the same theory holds true for several levels of parcels from baryons to stars in the cosmos: specifically, at the levels of nuclear force, van der Waals force, surface tension, and the force of gravity. A higher order of analysis clarifies other asymmetric ratios related to the halo structure seen in atoms and amino acids. We will also show that our minimum hypercycle theory for explaining the morphogenetic cycle (Naitoh, Artificial Life Robotics, 2008) reveals other temporal cycles such as those of economic systems and the circadian clock as well as the fundamental neural network pattern (topological pattern). Finally, a universal equation describing the spatiotemporal structure of several systems will be derived, which also leads to a general concept of quasi-stability.

1. Introduction

There are some mysterious fractal features common to the levels of various systems ranging from subatomic to cosmic.

First, let us consider a spatial aspect related to the golden and silver ratios. A neutron impacting uranium 235 produces smaller child atoms that often have an asymmetric weight ratio of about 2:3 between the golden and silver ratios. In contrast, varying the impact speed of neutrons results in a nearly symmetric division of uranium 235. [1] Purines and pyrimidines in biological base pairs, biological cells after divisions, liquid droplets, and stars in the cosmos also have a fusion of symmetry of 1:1 and asymmetry of around 2:3. [2-5] There are also mesons with "two" quarks and baryons with "three". [6] A model we developed previously based on fluid dynamics has revealed the reason for the fusion of symmetry of 1:1 and asymmetry of around 2:3 in the fractals found in nature.[2-5] The fusion of symmetry and asymmetry appears in various systems from atoms to stars, because each system is commonly related to a "breakup of flexible particles deformed". Then, a higher order of analysis clarifies the reason why size ratios over 2.0 are also seen in other biological molecules such as amino acids as well as liquid droplets. [7, 8]

Second is a rhythmical aspect in time. We can see several chemical oscillation phenomena in the bio-systems related to the origin of life, bacteria, morphogenetic processes, the brain, and the economy of the human network. The Brusselator, circadian clock, and heartbeat are also well-known examples. [9, 10] Until recently, little was known about the relation between these oscillations.

We have revealed that the core engine for giving life to living beings is a system of four categories of biological molecules, consisting of two types of information molecules and two types of functional molecules and resembling a four-stroke engine. [4] We have previously showed how the origin of life can be explained by this four-stroke engine, representing a minimum hypercyclic chemical reaction system. [4] This four-stroke engine induces the exponential growth of bio-systems.

Published under licence by IOP Publishing Ltd

Subsequently, fifth and sixth categories of molecules for repressing DNA replication were added to the four-stroke engine, resulting in several oscillating phenomena with chemical reactions. [11-13] A mathematical model based on these six categories of molecular groups showed that the oscillation cycle has a rhythm of about "seven" times the fundamental beat. [11-13]

This model clarifies the common principle underlying several levels of bio-chemical oscillations, because new organs emerge at "sevenfold" divisions of stem cells, while the circadian clock of about 24 hours is seven times the fundamental clock of about 3.5 hours. It is also stressed that one week has "seven" days. Several economic cycles are seven times the fundamental durable periods of certain products.

This interdisciplinary research spanning several categories of natural systems is expected to be effective, because we cannot see subatomic particles, the origin of the cosmos, or the origin of life, nor do we possess sufficient data on living beings and economical process.

However, little is known about the relation between spatial and temporal features. Thus, in this paper we will show a universal equation describing the spatiotemporal structure of various systems along with a generalized concept of quasi-stability.

* This workshop was supported in part by the Grant-in-Aid for the Global COE Program "The Next Generation of Physics, Spun from Universality and Emergence" from the Ministry of Education, Culture, Sports, Science and Technology (MEXT) of Japan.

2. Spatial structure

2.1. Model [2-5]

Here, we define a parcel as a flexible spheroid having two long and short radii of $a(t)$ and $b(t)$ dependent on time t, for the aggregation of neutrons and protons in each child atom resulting from the fission of a uranium 235 atom, a nitrogenous base in biological base-pairs of nucleic acids hydrated with a lot of water molecules, a biological cell just before division, and a star at breakup in the cosmos. The parcel becomes a sphere of the radius $r_d (=[ab^2]^{1/3})$ under an equilibrium condition. The deformation rate $\gamma(t)$ is defined as $a(t)/b(t)$, while a sphere without deformation corresponds to $\gamma = 1$.

Then, we consider the form of two spheroid parcels connected in line at the time of the breakup processes of uranium 235, at the replication stage of biological base-pair, at cell division, and at division of star. This is because atoms, nitrogenous bases, cells, and stars will have the shape of a gourd with two kinks at the later stage of breakup or at the first stage of coalescence. We derive a theory for describing the deformation and motions of the two connected spheroid parcels having two radii of r_{d1} and r_{d2} under equilibrium conditions and two deformation rates of γ_k $[k = 1, 2]$, while the size ratio of the two parcels is defined by $\varepsilon = r_{d1}/r_{d2}$.

We model the relative motion between the two parcels, nonlinear convections inside the parcels, and the interfacial force at the parcel surface. The interfacial force is evaluated in the form of σ / r^m where m and σ are constants and r is the curvature of parcel surface. Several types of forces such as nuclear force, van der Waals force, surface tension, coulomb force, and gravity can be explained by varying m. The relation $m = 1$ implies the surface tension of liquid. The mean density of the parcels is ρ_L.

We assume that the convection flow inside parcel is irrotational, i.e., potential one. There are random collisions of water molecules and electrons with the parcel such as biological molecules and cells. It is stressed that these random collisions from the outer region induce potential flow inside the flexible continuum particle, i.e., irrotational flow, because the fluctuations of impulsive starts and stops generate potential flow. [5] This potential flow is also applicable, because fluctuations entering the

Journal of Physics: Conference Series **344** (2012) 012008 doi:10.1088/1742-6596/344/1/012008

parcels such as bases, cells, and atoms will be close to those of thermal fluctuations, which are less dissipative. (Fluctuation dissipation theorem) It is well-known that the potential flow assumption is also applicable for liquid droplets. [2, 3] The potential assumption will be no problems also for stars, because of large size and high speed.

Moreover, we must consider that a parcel is not often a continuum, because the number of nucleons and water molecules inside the parcels for atom and nitrogenous base will be less than the order of 1,000. The scale for averaging, i.e., the minimum scale representing the phenomenon, will be smaller than that in continuum mechanics. Thus, this small averaging window leads to a weak indeterminacy of physical quantities such as deformation and density because of discontinuity of nucleons and molecules.

Here, we derive the relation between dimensionless deformation rate $\gamma_k(\equiv a_k / b_k\ [k=1,2])$ of each parcel dependent on dimensionless time $\bar{t}_k = \sqrt{\frac{8\sigma}{\rho_L r_{dk}^{2+m}}}\,t\ \ [k=1,2]$ and the size ratio of the two parcels of $\varepsilon = r_{d1}/r_{d2}$.

The stochastic governing equation having indeterminacy can be described as

$$\frac{d^2}{d\bar{t}_i^{\,2}}\gamma_i = \{m_{ci}(\frac{d}{d\bar{t}_i}\gamma_i)^2 + m_{cj}(\frac{d}{d\bar{t}_j}\gamma_j)^2 + m_{si}\,\gamma_i^{\frac{5}{3}-\frac{2}{3}m} + m_{sj}\,\gamma_j^{\frac{5}{3}-\frac{2}{3}m}\}/Det + \delta_{st}$$

$$[\text{for } i=1,2.\ j=1,2.\ i \neq j] \qquad (1)$$

with

$$m_{ci} = [(-\varepsilon - \varepsilon^4 + \frac{2}{3}\varepsilon E_{0j}\gamma_j^{-1/3})B_{0i} + \frac{2}{9}\varepsilon^{2+m}E_{0i}\varepsilon\gamma_i^{-4/3}]$$

$$m_{cj} = [\frac{2}{3}\varepsilon^{2+m}E_{0i}\gamma_j^{-1/3}B_{0j} - \frac{2}{9}\varepsilon^{2+m}E_{0i}\gamma_j^{-4/3}]$$

$$m_{si} = (-\varepsilon - \varepsilon^4 + \frac{2}{3}\varepsilon E_{0j}\gamma_j^{-1/3})C_{0i}$$

$$m_{sj} = \frac{2}{3}\varepsilon^{2+m}E_{0i}\gamma_j^{-1/3}C_{0j}$$

$$Det = -\varepsilon - \varepsilon^4 + \frac{2}{3}\varepsilon^4 E_{0i}\gamma_i^{-1/3} + \frac{2}{3}\varepsilon\ E_{0j}\gamma_j^{-1/3},\ B_{0k} = \frac{1}{3\gamma_k}\frac{\gamma_k^{\ 2}-2}{\gamma_k^{\ 2}-1/2},$$

$$C_{0k} = \frac{3}{8}\frac{2\gamma_k^{\ 2m} - 1/\gamma_k^{\ m} - \gamma_k^{\ m}}{\gamma_k^{\ 2} - 1/2},\ \text{and}\ E_{0k} = 3\frac{\gamma_k^{\ 7/3}}{\gamma_k^{\ 2}-1/2}[\text{for } k=1,2]$$

where the parameter δ_{st} denotes random fluctuation. It is stressed that this system is not the simple two-body problem of rigid body, because of flexible nonlinear deformations of the parcels.

[Equation 1 is derived only by the above assumptions and also purely mathematical transformation. The long derivation of Eq. (1) is in Ref. 3 confirmed by the referees, although only the stochastic term δ_{st} is not in Ref. 3.]

We then define the deviation from a sphere as y_i, which is equal to $\gamma_i - 1$. Taking the first order of approximation in the Taylor series leads to

$$\frac{d^2 y_i}{d\bar{t}_i^{\,2}} = [-\frac{2}{3}(3-\varepsilon^3-2\varepsilon^{2+m})\left(\frac{dy_i}{d\bar{t}_i}\right)^2 + 3(3-\varepsilon^3)my_i - 4\varepsilon^{1+m}\left(\frac{dy_j}{d\bar{t}_j}\right)^2 + 12\varepsilon^{1+m}my_j]/[3(\varepsilon^3+1)]$$
$$+\delta'_{st}, \qquad (2)$$

where the parameter δ'_{st} denotes random fluctuation.

Existing experimental data on breakup processes of liquid droplets of water and oil and biological cells show spheroid shapes at the timing of breakup or the later stage of breakup, which is the rate-determining stage. This also leads to a possibility that, for several levels of parcel breakups from subatomic to cosmic ones, the parcels will take approximately spheroid shape at the later stage after deformation. Thus, the present model reveals the essential principle underlying the fission process, biological divisions, and stars, whereas multi-dimensional calculations using supercomputers shows the process in detail.

2.2. *Size (representative length)*

Equation 2 shows that a symmetric ratio of 1.0 (ε=1) makes the first term on the right-hand side of the equation zero, while an asymmetric ratio of $\sqrt[3]{3}$ around 1.5 (ε^3=3) makes the second term zero for each m. The size ratios of 1.00 and approximately 1.50 can be described by the unified number of the n-th root of n.

We define a system as being quasi-stable when only one term on the right-hand side of the differential equation system governing the phenomenon is zero. (The quasi-stable principle may also be defined by the condition that either of volume or surface forces is zero.) The system of two parcels connected is relatively quasi-stable because d^2x/dt^2 becomes smaller when the size ratio of connected parcels takes the values of ε=1 or ε^3=3.

The concept of quasi-stability is weaker than neutral stability. [3, 4, 5] The quasi-stability concept is necessary for living beings, because stronger stability cannot bring variations, i.e., adaption for environmental change and evolution. This quasi-stability is also possible for nonliving systems such as atoms, because atomic systems also vary in a long time. (This is also because both atomic systems and living beings have dense interaction between particles in closed volumes.)

Next, let us look at biological systems containing water flows. We can classify the five bases of adenines (A), guanines (G), cytosines (C), thymines (T), and uracils (U) into two groups: purines and pyrimidines. Purines, i.e., A and G, have a relatively large size, while pyrimidines, i.e., C, T, and U, are small. Asymmetric base pairs such as the Watson-Crick type of about 1: $\sqrt[3]{3}$ are used in living beings. This grouping specifically refers to the asymmetric size ratio of purines and pyrimidines of around 1.50 in their hydrogen bonds within DNA and RNA, although a symmetric size ratio of 1.00 is often observed in RNA. [4, 5] Symmetric and asymmetric size ratios are also observed at the cell level of microorganisms such as yeast. [2, 3]

These ratios of 1:1 and about 1:1.5 also correspond to those of child atoms generated by the breakup of uranium 235. This means that the probabilities of the size ratios of 1:1 and about 2:3 are relatively high, because uranium will have the shape of a gourd with two kinks just before its breakup.

The quasi-stable ratios of 1: $\sqrt[n]{n}$ for n=1 and 3 appear for each m (see Eq. 2). This universality also leads to the possibility that the present model can be applied for several levels of parcels from baryons to stars in the cosmos [5, 6, 14]: specifically, at the level of nuclear force, coulomb force, van der Waals force, surface tension, and force of gravity.

As Eqs. 1 and 2 show a slightly vague solution for the phenomenon, this indeterminacy also implies that size variations of ε are possible in a limited range. This indeterminacy permits the possibility of sizes around 1:1 and also around 1: $\sqrt[3]{3}$, i.e., between 1:1 and about 2:3, although 1:1 and about 2:3 are more.

Let us consider the reason why the energy conservation law and variation principle such as the Bohr model do not explain the fusion of symmetry of 1:1 and asymmetry of around 2:3 in size and density (number density). The reason is that the momentum conservation in Eq. 1 models all of the relative motion between two parcels, nonlinear convections inside the parcels, interfacial forces at the parcel surfaces, and collisions with smaller molecules, whereas previous models based on energy conservation have tended to eliminate the relative motion between parcels. A second reason is that the quasi-stability principle was proposed instead of the variation principle.

Another important point is that natural systems should be explained in terms of four conservation concepts regarding mass, momentum, moment, and energy, as is seen in the thermo-fluid dynamics. Conservative quantity is not only energy. The present new explanation for two particle systems is necessary, because explanations based on two conservativity principles such as momentum and energy should be done. The following section also shows the importance of mass conservation law.

[The effect of special theory of relativity between mass and energy will not influence for the size ratio of child atoms obtained after the fission process of uranium 235 very much, because the effect work on both two child atoms at an identical rate. Moreover, impact of only one neutron to uranium will not produce fast deformations of uranium at the later stage of fission.]

2.3. Number ratio

It is also well known that several atoms in nature have the number ratios of protons and neutrons between 1:1 and 2:3. Here, let us examine the reason why larger atoms have larger number ratios close to 2:3.

The first and second terms on the right-hand side of Eq. 2 also show that symmetric (ε=1) and asymmetric (ε^3=3) parcel divisions are more stable for small disturbances of motion (dy/dt) and parcel deformation y, respectively (Fig. 1).

Let us separate baryon aggregation into two parts: the internal side around the center of the aggregation and the external side close to the surface. External baryons close to the surface move relatively easily, because one part of the baryon is free without any connection to other baryons. However, internal baryons often receive forces from many directions due to the presence of other ones, making it relatively difficult for them to move relative to the origin on the earth. Thus, inner baryons deform relatively easily without any translational motion of the gravity center. As a result, the inner and outer baryons determine whether baryons are asymmetric or symmetric, respectively. [4, 5] An important point is that the concept of inner asymmetry and outer symmetry can be seen.

Larger aggregations of parcels such as Thorium (Th) including baryons more than Helium (He) have more inner baryons, because the surface/volume ratio of the aggregation becomes smaller as size increases. (Fig. 2) More inner baryons for larger atoms bring more asymmetric number ratio of protons and neutrons, as larger nucleic acids such as rRNA also have more asymmetric number ratios of purines and pyrimidines close to 2:3. (Fig. 3) This is because of mass conservation law, i.e., because heavier particles such as purines and protons can be generated with less numbers. [4, 5]

A mysterious thing is that the masses of stable proton and neutron are almost same, while child atoms generated by fission of uranium 235 and nitrogenous bases (pyrimidine and purine) have the different weight ratio around 2:3. Qualitatively, proton is a little heavier than neutron. Thus, relatively heavy protons can be with less number in atomic core, because of mass conservation law. We may quantitatively be able to explain this mystery by the fact that the mass ratio of proton and neutron are around 2:3 as the forms such as lambda particle before atomic cores are stabilized, because the up and down quarks have the weight ratio between 1:1 and 1:2. Thus, the later stage of fission process for

Journal of Physics: Conference Series **344** (2012) 012008 doi:10.1088/1742-6596/344/1/012008

generating atomic cores is described by the present theory. Then, the final or post-fission stage generates the atomic cores of stable protons and neutrons. This explanation is possible by the analogy with biological system, because nitrogenous bases are also connected with heavy molecules such as ribose or deoxyribose, which lead to nucleic acids. The weight ratio of purine- deoxyribose and pyrimidine- deoxyribose is close to 1:1, i.e., symmetric value. (Fig. 4)

The present theory also indicates that the number ratios around 1:1.3, around the center between 1:1 and about 1:1.5, will be relatively unstable. (Fig. 3) This can be understood by the fact that both fluctuations of deformation and velocity enter into the parcels connected with the medium number ratios of about 1:1.3, whereas only velocity fluctuations enter into small parcels such as those of Helium (He) and only deformations occur for large atoms such as Thorium (Th). Actually, atoms, which are a little larger than lead (Pb), are unstable.

Atoms larger over 300 nucleons are unstable, which cannot have the size ratios such as 1:1 and about 2:3. This can be explained by the fact that larger deformations enter into the atoms, because larger atoms get larger deformations.

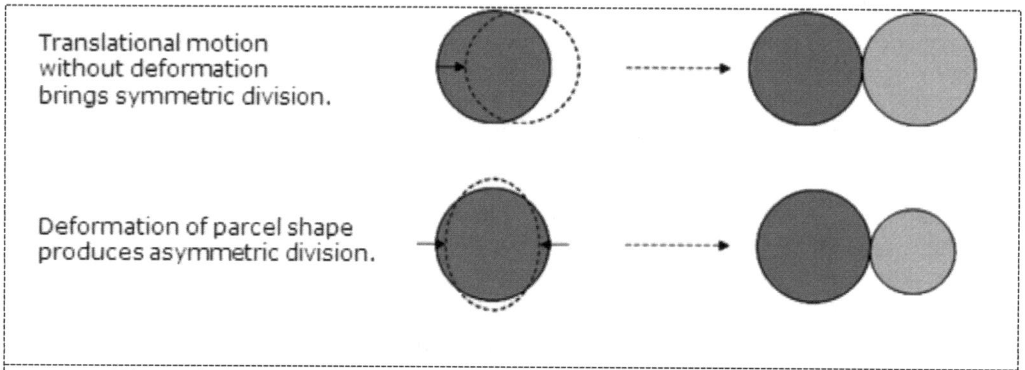

Figure 1 Asymmetric division due to parcel deformation and symmetric division in the absence of deformation.

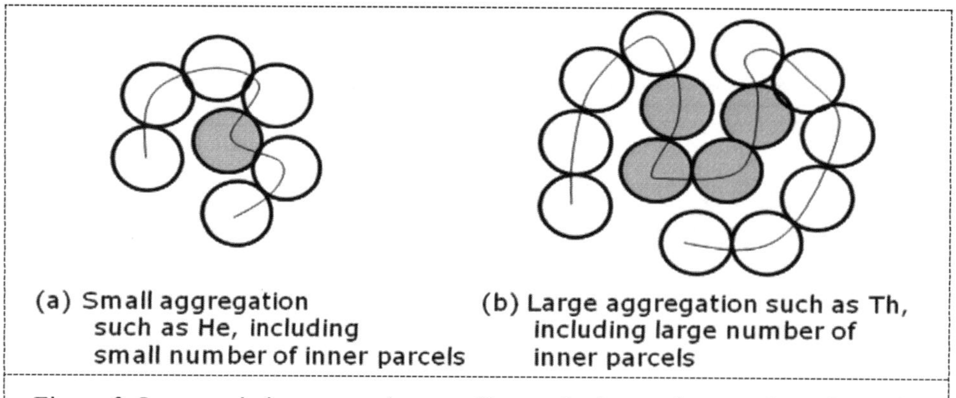

Figure 2. Increase in inner parcels according to the increasing number of parcels.

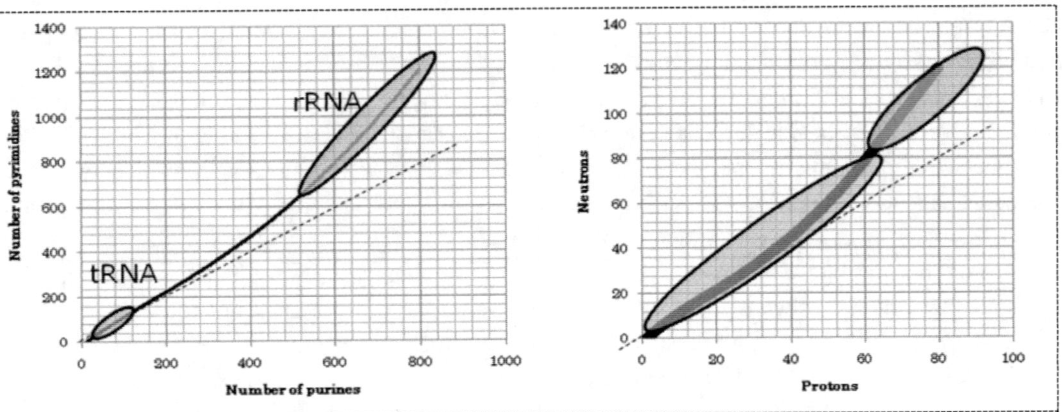

Figure 3. Density ratio of purines and pyrimidines in RNAs, which is similar to the density ratio of protons and neutrons in stable atom core. (mRNAs are between tRNAs and rRNAs, which are relatively unstable.) Left: RNAs, Right: Atoms

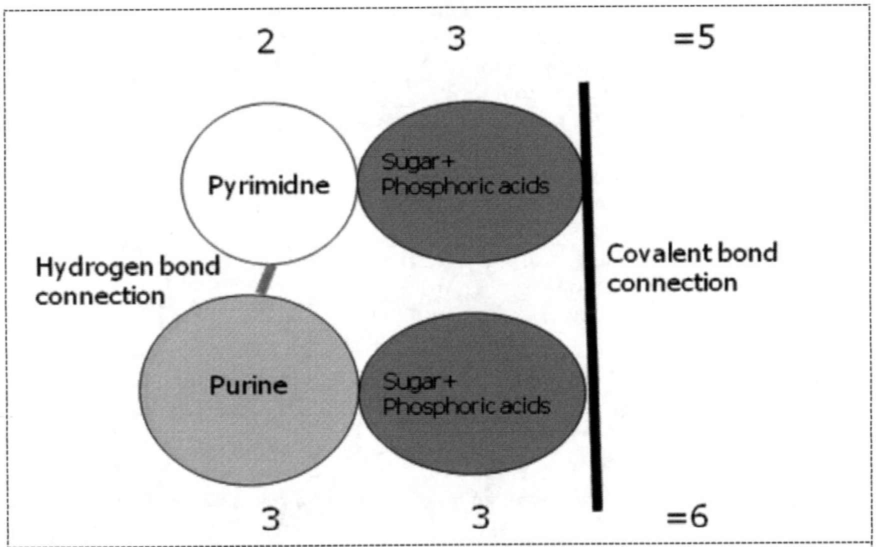

Figure 4. Pyrimidine and purine in Watson-Crick base pairs connected with sugars by covalent bonds.

Thus, this theory explains the reason why relative heavy atoms have more asymmetric number ratios close to 2:3 and also why the atoms having the number ratios around 1:1.3 are relatively unstable, while the energy conservation law such as those of Bohr clarifies the inevitability that neutron-rich system is more stable than proton-rich one. [The quasi-stability concept applied to momentum conservation law and mass conservation law bring the result compatible with that the Bohr's drop model based on the variation principle for energy conservation law and also gives a further physics underlying the Shell model related to the magic number.]

2.4. Taylor series

Next, we take a higher order of the Taylor series for Eq. 1. Even-numbered terms such as the second and fourth ones show no other quasi-stable size ratios. However, odd-numbered terms result in other quasi-stable ratios. The third term in the Taylor series results in a quasistable ratio of about 3.5, the fifth term in ratios of about 2.5 and 2.1, and the seventh in a ratio of about 1.78. An important point is that the terms of orders higher than the ninth are absolutely unstable, which leads to the existence of the maximum size limit of atoms. (Table 1)

It is also stressed that liquid fuel droplets generated by injectors and child atoms broken up from the fission of uranium 235 also have a threefold variation of sizes at the maximum and also that the molecular weights of the twenty types of amino acids show a threefold variation between 240 of cysteine as the maximum and 75 of glycine as the minimum. [1, 4, 5] The higher-order of analysis clarifies the ratios over 2:3 in several systems. (Table 1)

Free atoms in atmosphere and molecules outside living beings are not with the special ratios such as 1:1, about 2:3, and 3.5, because free atoms in atmosphere and molecules outside cells are relatively difficult to interact each other, while subatomic particles such as baryons and biological molecules like bases, nucleic acids, amino acids, proteins are strongly interactional insider closed regions.

Table 1 Quasi-stable ratios observed in atoms and biological systems.

n-th term in Taylor series	Quasi-stable ratio	Atoms	Biological molecules
1	1:1 & about 1.1.5	Main size ratios of child atoms broken from uranium 235. Number ratios of neutrons and protons in atoms	Nitrogenous bases in DNAs and RNAs (Purines:Pyrimidines). DNA polymerases I, II, and III (I:II, I:III)
3	about 1:3.5	Higher limit of size ratios of child atoms broken from uranium. Number ratio of protons and neutrons in He 10	Amino acids
5	about 1:2.5 & about 1:2.1	The intermediate size ratios of child atoms broken from uranium. Number ratio of protons and neutrons in Mg 32	rRNA (23S:16S)
7	about 1:1.78	The intermediate size ratios of child atoms broken from uranium.	Ribosome (50S:30S)
9, 11	unstable	-	-

2.5. Further possibilities

Our previous report based on the quasi-stability concept applied to momentum conservation [2-5, 7, 8] revealed the reason why several particles such as biological cells, nitrogenous bases, and liquid droplets have the bimodal size ratios of about 2:3 and 1:1. There are mesons with “two” quarks and baryons with “three”. Therefore, an analysis based on Eq. 1 may also clarify the ratios in the elemental particles such as quarks.

The present paper extended with stochastic mechanics and indeterminacy principle and the higher order of analyses also reveal the other ratios over 2:3. The halo structures such as H10 and M32 have the number ratios of neutrons and protons over 2:3. [16] These number ratios similar to those in largest and smallest child atoms broken from uranium 235 and biological molecules such as amino acids will also be explained by the present theory of the higher-order of accuracy.

3. Temporal structure

3.1. Minimum drive element for the origin of life (Fore-stroke engine of biomolecules) [4]

Journal of Physics: Conference Series **344** (2012) 012008 doi:10.1088/1742-6596/344/1/012008

Both information strings such as DNA and catalyers such as enzymes are necessary for bio-systems, which can replicate to counter the natural degradation of biomolecules. At least two types of information strings for the origin of life are necessary, which code catalysers for generating information strings and those for generating catalysers. This also implies the inevitability of two catalysers. Thus, the minimum system for self-organization at the origin of life has four types of molecules: two information strings and two catalysers.

We call molecules having information such as DNA "infostrings", while the catalysers such as enzymes are called "catalysts". The core system shown in Fig. 5 includes four molecules consisting of Infostring 1 that codes the catalyst generating infostrings, Infostring 2 that codes the catalyst for generating catalysts, Catalyst 1 generated by Infostring 1, and Catalyst 2 produced by Infostring 2. This four-cycle system works as a closed loop, if elements such as nTP and amino acids are input from the outside with energy. This is the minimum 4-stroke engine (minimum hypercycle), which produces molecules exponentially in cases where there are no degradations.

We define the densities of Infostrings 1 and 2 as D1 and D2, respectively. Then, the densities of Catalysts 1 and 2 are defined as R1 and R2, respectively. This yields the following essential equation system.

$$D_i^{N+1} - D_i^N = \alpha_{i1} D_i^N \otimes R_1^N$$
$$R_i^{N+1} - R_i^N = \alpha_{i2} D_i^N \otimes R_2^N \qquad (3a, 3b)$$

where $D_i \otimes R_j$ denotes the smaller value among D_i and R_j with i=1, 2 and j=1, 2. Then, α_{ij} and N imply an arbitrary constant and generation, respectively.

Equation 3 can also be written by the form of differential equation as

$$\frac{d}{dt} D_i = \alpha_{i1} D_i \otimes R_1$$
$$\frac{d}{dt} R_i = \alpha_{i2} D_i \otimes R_2 \qquad (4a, 4b)$$

Results computed with this equation system show nearly exponential increases in Di and Ri. It is stressed that variable R1 on the left-hand side of Eq. (3b) never appears on the right-hand side of the equation, although common variables can be observed on both sides of Eq. (3a). Equation (3b) essentially shows the asymmetry topologically.

The simplest hypercycle of D1, D2, R1 and R2 works as "the engine of life" for promoting more complex reaction networks.

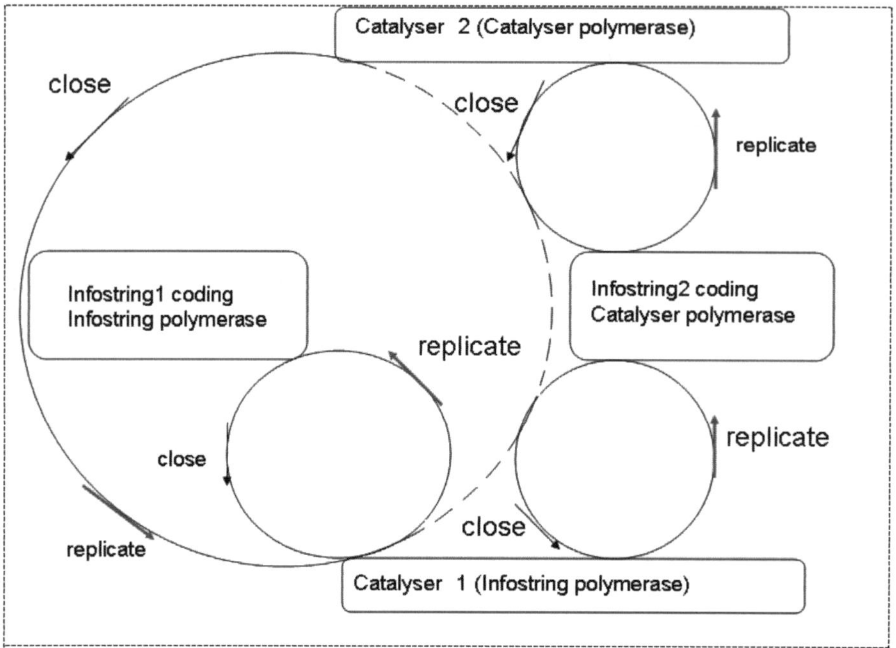

Figure 5. Four-stroke engine as a minimum hypercycle.

3.2. Minimum oscillator for morphogenetic process (Six-stroke engine of bio-molecules) [11-13]

Let us consider bacteria, eukaryotes, and cells in higher-order living systems. Four types of fundamental molecules (D1, D2, R1, and R2) must be extended to four molecular groups, because life today uses numerous types of genes and proteins. We can then define four macroscopic groups as follows.

D1: Gene group 1 that codes enzyme system R1 for DNA production
D2: Gene group 2 that codes enzyme system R2 for enzyme production
R1: Enzyme system 1 for DNA production, including DNA polymerases, topoisomerases, helicases, ligases, and primases
R2: Enzyme system 2 for enzyme production, including ribosomes with rRNAs and ribosome proteins, RNA polymerases, mRNAs, and tRNAs

It is stressed that D2 and R2 are for activating reactions, as is shown in the previous section. Next, we define D3 and R3 as other molecular groups for repressing reactions as in the case of transcription factors. These two groups are incorporated in the four groups of D1, D2, R1, and R2, because today's cells and the morphogenetic processes of multi-cellar systems use negative controllers such as Oct-4 and SOX2 for producing tissues and organs [17]. Accordingly, D3 and R3 can be defined as follows.

D3: Gene group 3 that codes protein system R3 for repressing Di gene expressions
R3: Protein system 3 for repressing Di gene expressions, including the proteins of the transcription factors such as those of Oct-4 and SOX2

This leads to a macroscopic model having six types of molecular groups, or in other words, a six-stroke engine. (Fig. 6) We can describe the densities of the six groups at generation N after the mother cell generation by the following equations.

$$D_i^{N+1} - D_i^N = \alpha_{i1} D_i^N \otimes R_1^N \quad ,(i = 1,2,3) \qquad \text{(5a, 5b)}$$

$$R_i^{N+1} - R_i^N = \alpha_{i2}\delta(D_i^N - \varepsilon_i R_3^N) \otimes R_2^N \quad ,(i = 1,2,3)$$

where $D_i \otimes R_j$ denotes the smaller value among D_i and R_j with i=1, 2, 3 and j=1, 2, 3 and also where $\delta(x)$ denotes the larger one of x or 0, i.e., max (x, 0). Parameter ε_i with i=1, 2, 3 denotes an arbitrary constant for controlling repression rates. In this report, we set to be $\varepsilon_1 = \varepsilon_3 = 0$, while $\varepsilon_2 = 1.0$, because R3 usually represses D2.

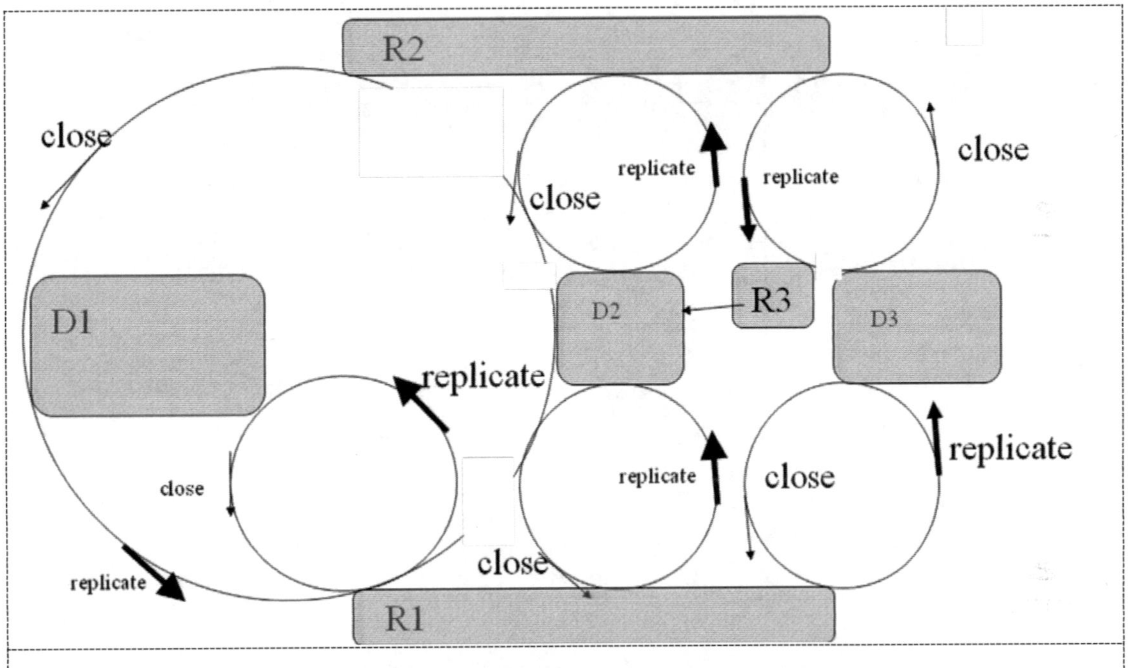

Figure 6 Six-stroke engine as a minimum clock.

Figure 7 obtained by solving Eq. 5 shows the density ratios of Di and Ri. The condition of D2/R3 > 1.0 means that a part of D2 is not covered by R3. The condition of D2/R3 < 1.0 means that D2 is completely blocked by redundant R3. When R3 is dense in stem or induced pluripotent stem (iPS) cells, it means that stem and iPS cells can be reprogrammed by the presence of much Oct-4. This oscillation of D2/R3 will lead to changes in the gene combination for expressions, i.e., the emergence of new organs. Figure 7 clearly shows that the antagonism between the negative controller R3 and the positive replication factors R1 and R2 induces bifurcation events at rhythmic intervals constituting about seven divisions, although the intervals are slightly chaotic and the vibrational amplitude is attenuated.

The mathematical model based on these six categories of molecular groups, Eq. 5, showed that the oscillation cycle has a rhythm of "seven" times the fundamental beat. [11-13] It is stressed that this cycle of about seven divisions, i.e., the branching time between periodic bifurcation events, corresponds to the emergence timing of blast cysts, embryos, germ layers, tissues, and organs, which can be observed for about every seven cell divisions. Emphasis is placed on the fact that three system parameters, α_{i2}, have less influence on the time cycle. The cycle of about seven divisions is fairly stable. (The time cycles are between four and nine beats, while varying α_{i2} between 0.0 and 5.0. The cycle is about sevenfold beat on the average.)

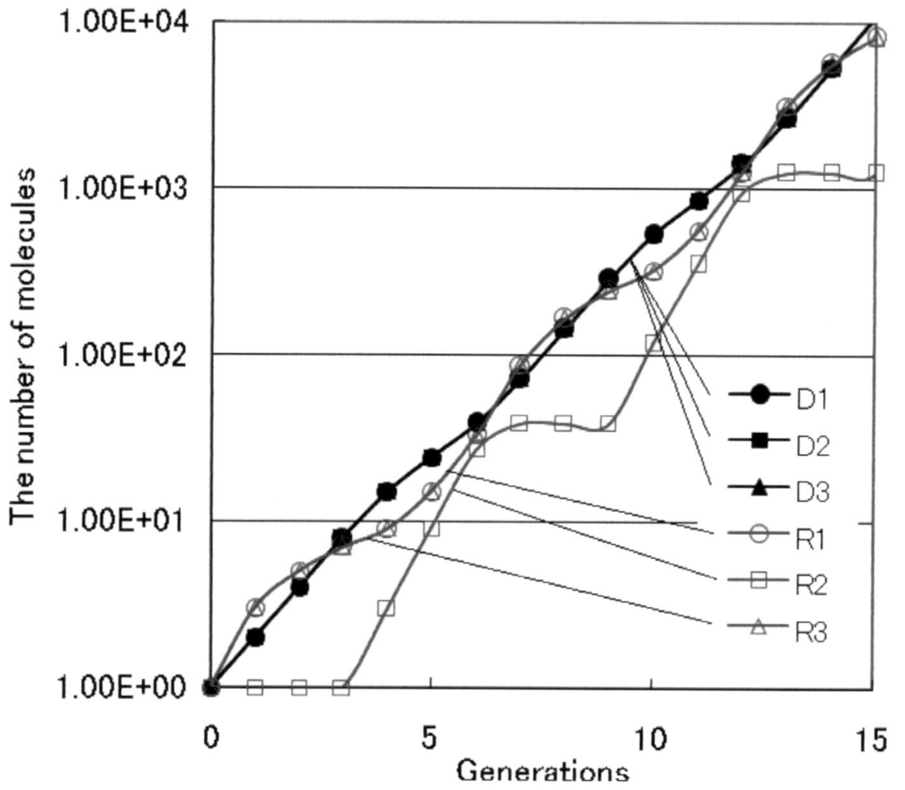

Figure 7 Time histories of Di and Ri during 15 generations , with the initial conditions of Di = 1 and Ri = 1.

3.3. Economic cycles and human society

There are several types of boom-and-bust economic cycles: the Kondratev cycle of about 50 years, the Juglar cycle of about 7 years, and the Kitchin cycle of about 3.5 years, which are observed in human network systems. [18, 19]

The shortest economic cycle is one week of "seven" days. The longest repetitive catastrophes occur about every 70 years, which is seven times the actual durable period of automobiles of about 10 years. Thus, we hypothesize that several types of economic cycles are about seven times the length of the fundamental production cycles (durable periods). New organs that emerge with a sevenfold rhythm may correspond mathematically to the rise of new production or new business categories. [We classify all things in our economic system into two major categories: information and functional objects. Computers and industrial robots are examples of functional objects. Functional objects can be further divided into two subgroups: those such as computers for generating information and those such as industrial robots for generating functional products. Information can also be classified into two subgroups: for designing information machines such as computer and for functional products. These four subgroups correspond to D1, D2, R1, and R2 in Eqs.(3) and (5). Then, fifth subgroup of R3 such as alcohol for depressing D2 and sixth subgroup of D3 for generating R3 lead to the cycles.]

Journal of Physics: Conference Series **344** (2012) 012008 doi:10.1088/1742-6596/344/1/012008

3.4. Fundamental neural network pattern

Next, let us remind the circadian clock in the brain, having a cycle of 24-25 hours. We know that, when suprachiasmatic nuclei (SCN) in brain are removed, the clock shows a cycle of about 3-4 hours. [20] This means that the sevenfold-beat cycle based on the six-stroke engine of molecules will also be present in the brain. The fundamental topological pattern of neural network in brain may also be that in Fig. 6, which is similar to Fig. 8. Six variables of Di and Ri (i=1-3) are redefined as activation level of neurons.

Neurons D1, D2, D3: Provoked pulse induced by input from outer region to Di goes into R1, which activates Di cyclically.
Neuron R2: Provoked pulse induced by input from outer region to R2 goes into D2, which activates R2 as accelerator.
Neuron R1: Provoked pulse induced by input to from outer region to R2 goes into D1, which activates R1 as accelerator.
Neuron R3: Provoked pulse induced by input from outer region to R2 goes into D3, which activates R3 as depression neuron.

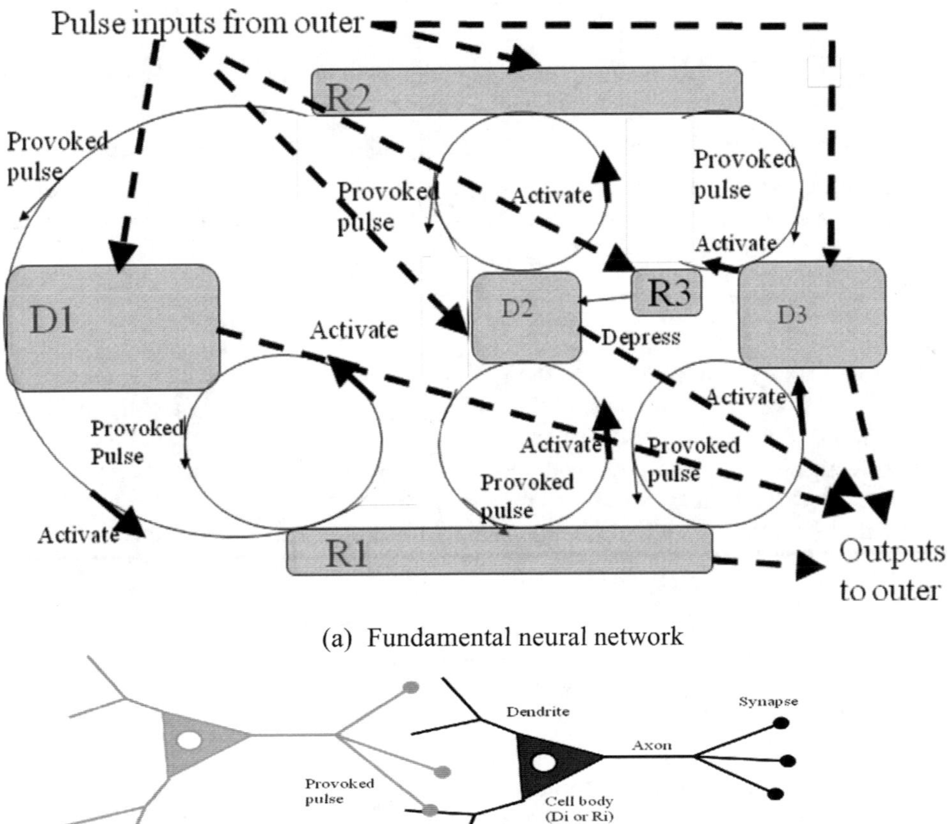

(a) Fundamental neural network

(b) Local image of Di and Ri

Figure 8 Fundamental topology of neural network.

Several systems and functions inside brain such as memory, thinking, pyramidal area, emotion, language, visual and auditory senses may also be governed by this seven-beat cycle of Fig. 8. It is also stressed that the fundamental neural network pattern inside brain in space (topological pattern) will also be dominated with a fusion of symmetric and asymmetric connections in Fig. 8.

4. Quasi-stability extended to spatiotemporal structure

Let us think about whether or not the quasi-stability concept can also be applied to the evolutionary process. The macroscopic kinetic model describing the time-dependent populations of the four main groups of species in the evolutionary process [21] shows that only one group of species does not increase at the time of a transition in evolution, while the other groups vary in time. [Figs. 5 and 17 in reference 21]

Moreover, Eq. 5 explaining the biological cycles occurring in the brain and morphogenetic processes and economic cycles [11-13] shows that, when only one variable among six is invalid, drastic transitions occur in morphogenetic and economic processes. (Fig. 7)

Thus, a quasi-stable condition, i.e., a weakly stable situation at the transitions such as fission, evolution, and the emergence of organs, can be redefined by the rule that only a part of the equation system, i.e., only one term or one variable, is invalid at the time of a transition.

5. Governing equation unified for both spatial and temporal structures

Equation 4 for D_1 can be transformed to the form of

$$\frac{d^2}{dt^2}D_1 = m_c[\frac{d}{dt}D_1]^2 + m_s D_1 + m_x \tag{6}$$

with the additional term of m_x. The result is essentially identical to Eq. 2 and also to the deterministic equation for evolution in Reference 21.

Here, we eliminate only the third term m_x on the right hand side of Eq. 6 to obtain

$$\frac{d^2}{dt^2}D_1 = m_c[\frac{d}{dt}D_i]^2 + m_s D_1 \quad . \tag{7}$$

Equation 7 is essentially nonlinear due to the first term on the right-hand side. However, this equation has an analytical solution. [22] Increasing the initial disturbance leads to a transition from a trajectory of oval oscillations to parabolic types. Then, variations of input energy bring two types of modes (oval or parabolic), which are stable sustainment such as information storage and drastic change typified by the emergence of new organs.

We can conclude that living and nonliving systems from atoms to stars employ two types of modes due to Eqs. 6 and 7, in order to use quasi-stable spatiotemporal structures between stable and reorganized conditions.

6. Four-dimensional scenario of the biological self-organizing processes: clarified by the fusion of the foregoing sections

6.1 Quasi-stability and neutral stability in Section 2

Let us think about the asymmetric pattern of the Karman vortex streets [23] generated downstream of a bluff body. Emphasis is placed on the fact that Karman's theoretical analysis based on the neutral stability principle clarifies only the instantaneous stability of the vortex array in space, but it does not show the temporal development of the vortices. Many researchers have confirmed that the analysis based on the neutral stability at the instantaneous stage has a high probability of showing the specific pattern of vortices over a long period in the natural observation.

Thus, this precedent set by Karman is further evidence that the quasi-stability principle, which is a little weaker than neutral stability, can explain why parcels are connected by biological size ratios such as 1:1, about 2:3 or about 1:3.5.

Journal of Physics: Conference Series **344** (2012) 012008 doi:10.1088/1742-6596/344/1/012008

Here, let us recall that the terms of orders higher than the ninth are absolutely unstable, which leads to the existence of the maximum size limit of parcels in Table 1. This also shows that, in order to execute four-dimensional numerical computations for the self-organizing process of living beings, the 9th order of accuracy is necessary. This is a very important guiding principle for near-future simulations on supercomputers.

6.2 Probability and population in Section 3

The 4- and 6-stroke engine models described in Section 3 are in the category of population dynamics, which reveal the probability of each molecular group. Various processes, including those of living beings, are stochastic because of the indeterminacy shown by Eq. 1 in Section 2. Thus, we should consider the probabilities at which bio-chemical reaction processes such as replication, breakup, and degradation occur with some mutations. The replication process of DNA and cell divisions are not deterministic.

A quasi-stability analysis tells us that, when a base pair of two parcels connected (a base pair hydrated) and a cell just before division have the shape of a gourd with two kinks such as what occur at the later stage of breakup or at the first stage of coalescence of two parcels, the probabilities are high for size ratios of 1:1 and about 2:3, while various size ratios of candidates of nitrogenous bases and amino acids are generated by coalescence and breakup with an energy input. (An example of breakup in living beings is the first stage of DNA replication, while an example of coalescence is the final stage of DNA replication.)

High probabilities of size ratios of 1:1 and about 2:3 stabilize the primitive information of base pairs, which induce the complex information of DNA. As a result, there is greater production of various enzymes with different functions, leading to the probabilities of stable cycles of 4- and 6-strokes shown above. The probability of the above-mentioned scenario for self-organization of today's human beings and ancient lives is larger than that due to a fully random strategy, although the probability of this scenario based on quasi-stability will not be very high.

Some amino acids softer than the bases may deform easily. This also leads to quasi-stable size ratios over 1.5 (about 1:3.5, 1:2.5, 1:2.1, and 1:1.78) for large deformations, because of the higher order terms of the Taylor expansion from the 3rd to the 7th, while size ratios of 1:1 and about 2:3 are also possible. Some large proteins are also very soft and may be unstable.

Thus, biological processes are not deterministic, nor do they have a one-to-one correspondence because of several passages of chemical reaction networks, which have the diversity of redundancy (many-and-many), degeneracy (many-to-one), and versatility (one-to-many). [24]

The foregoing discussion has concerned parcels connected by hydrogen bonds. Next, let us think about an interesting example of parcels connected by covalent bonds. Our previous report shows that the size distribution of various amino acids is similar to that of human chromosomes. [14] This similarity may imply that the present theoretical model can also be applied to the disconnection of covalent connections.

6.3 Oval and parabola in Section 5

The analytical solution for Eq. 7 in Section 5 shows that increasing the initial disturbance leads to a transition from a trajectory of oval oscillations to parabolic types. This transition implies that variation of the input energy brings two types of modes (oval or parabolic), which make possible both stable sustainment such as information storage in DNA and fast replication of DNA. This tendency of stability against a small disturbance and fast breakup due to a large disturbance is also confirmed by a numerical analysis of Eq. 1, when the size ratios of parcels are 1:1 and about 2:3. [3] This corresponds to the fact that a large energy input in the mitotic phase leads to fast breakup of DNA with base pairs having a size ratio of 1:1: and about 2:3 and more cells, while the size ratios are relatively stable for small disturbances during the silent period without any large energy input.

6.4 Four-dimensional scenario for information, structures, and functions of various biological molecules and organs

There is also a 2:3 ratio in immune globulin (IgX). Two types of large immune globulines (IgM and IgE) and three small ones (IgG, IgD, and IgA) are naturally selected, while the weight ratio of the small and large heavy chains is also 2:3. [11]

The other large molecule, ribosome, includes ribosomal RNA (rRNA). The average frequency ratio of purines and pyrimidines in lots of types of rRNA sequences is actually about between 1.0 and 1.5.

Both hydrophilic and hydrophobic amino acids exist in proteins. If we employ the classification of R-side chain grouping, the frequency ratios of hydrophilic and hydrophobic amino acids in proteins are also asymmetric in many cases, between 1:1 and approximately 2:3. [4] This asymmetric symbiosis of hydrophilic and hydrophobic molecules also produces the complex convexoconcave shapes of proteins, as hydrophobic parts are often folded inside proteins.

How are biological functions generated from the fundamental information represented by the asymmetric and symmetric ratios of 2:3 and 1:1 described in the previous sections?

Purine and pyrimidine pairs do not form in the presence of only one type of base. An extremely large frequency ratio of purine and pyrimidine, say, far larger than 1.5, cannot produce the stem in tRNA. The fact that tRNA has the weakly-asymmetric density ratio of purines and pyrimidines between 1:1 and about 2:3 produces the clover structure with loops (leaves) and stems, because the loci inside the leaves (loops) are free from hydrogen-bonded connections. [4] This clover structure is constructed with three leaves similar to fingers, which can catch other molecules and also carry the other molecules to ribosome. The "catch" is an important function for living being. As larger molecules such as rRNA and some enzymes are also with repeats of the leaves such as those in tRNA, the fingers catch mRNAs.

The above-mentioned asymmetric ratio around 2:3 reveals the relation between the three-dimensional structure and one-dimensional information, i.e., the reason why RNA such as tRNA and rRNA have clover structures with fingers, while DNA has a simple spiral. An important point is that the combination of a string (stem) and a ring (loop) in RNA results in a function, although a string such as DNA with information alone cannot generate a function.

Another interesting point is that the anti-codon inside the leaf of tRNA induced by the asymmetric ratio essentially permits a maximum of sixty-four patterns, because A, U, G, and C are possible for each of the three loci (4 x 4 x 4 = 64). Then, because the third locus inside the anti-codon has a relatively weak connection with the codon, the minimum number of patterns is sixteen. This implies that the number of amino acids is between sixteen and sixty-four. The weak connection at the third locus permits twenty types of amino acids, a little more than sixteen. It is emphasized that clarification of the leaf size with an anti-codon of three loci related to the asymmetric ratio of 2:3 for densities also leads to the general inevitability of twenty amino acids.

There is another important mechanism of the self-organizing process in living beings. It is well known that the electron cloud is not spherical around an atom heavier than helium (He). A non-spherical electron cloud induces string-like and ring-like molecules in amino acids and nitrogenous bases, without any spherical connections of carbons. Nitrogenous bases without a spherical electron cloud generate string-like DNA and RNA. This is an essential principle of chemistry.

Now, let us consider a rotating spheroid that is generated by deformation of a parcel such as a cell. The spheroid is geometrically asymmetric due to its rotating axis. The discrepancy from a complete sphere, i.e., its scabrous shape, induces a string-like connection of deformed parcels, although parcels with less deformation are connected as a spherical aggregation or a spherical surface aggregation in two- or three-dimensional space. Repeats of strings and spheres can also be observed in larger scales. [5, 14] Although oospheres still maintain a three-dimensional spherical shape overall during several divisions after being fertilized by antherozoids, the sphere gradually comes closer to being a one-dimensional string due to asymmetric divisions. The string will be close to the next larger sphere like yarn waste because of its flexibility. [5]

In our previous reports, we also simulated the morphogenetic process of the complex cerebral shape in the brain, including its main blood vessels, as well as that of asymmetric inner organs such as the liver and symmetric outer organs like the arms and legs. [4, 5, 25] These results correspond to the principle of inner asymmetry and outer symmetry discussed in Section 2.

7. Conclusion

A quasi-stability analysis shows that, when a pair of two nitrogenous bases, a cell just before division, an atom at the breakup has the shape of a gourd with two kinks such as what occur at the later stage of breakup or at the first stage of coalescence of two parcels, the probabilities are high for size ratios of 1:1, about 2:3, and also 1:3.5, while various size ratios of candidates of nitrogenous bases and amino acids are generated by coalescence and breakup with an energy input.

The population model of four and six strokes clarifies the standard circuit controlling various stages of living beings.

Fusion research based on time and space demonstrates a possible four-dimensional scenario of biological self-organization, although there are still many mysteries in natural systems.

Appendix: Volume and weight ratios

The weight and size (representative length) ratios of purines and pyrimidines are about 1.5. The asymmetric pair of flexible spheroid parcels of “purines surrounded by water molecules” and “pyrimidines surrounded by water molecules” will also have the size (length) ratio between 1.40-1.50. The weight and size (length) ratios of nitrogenous bases are proportional to the parcel size ratio (parcel length ratio). However, the weight and volume ratios of the two spheroid parcels are 3.0. The difference between the weight ratios of parcels and nitrogenous bases, i.e., about 1.5 for the weight ratio of bases and 3.0 for the weight ratio of parcels, is possible, because the molecular weight ratio of the parcels including water is larger than that of the purine-pyrimidine bases and also because the electric density distribution of Pai electrons will not be in a two-dimensional distribution, while the nitrogenous bases have one-dimensional string (ring).

Emphasis is placed on the fact that the size (length) and weight ratios of child atoms broken up from uranium 235 will also be close to 2:3, because atomic cores are mathematically similar to one-dimensional strings distorted or rings.

What is the difference between the weight ratio of two child atoms separated from uranium 235 and the weight ratio of the parcels including child atoms? It will be possible that the weight ratio of parcels is larger than that of protons and neutrons in atom or child atoms split from uranium 235, because of immerse mass. Immerse mass proportional to the parcel weight only appears while deformation and breakup occur. The immerse mass can be understood analogically from the fluid dynamic effect in which impulsive deformation motion is accompanied with added mass [15]. This immerse mass produces a weight ratio of 3.0 for the parcels, although the weight ratio of child atoms is about 1.5. This mass may be related to gluon, quark condensation, or the relativity theory effect. Parcels smaller than cells may be with immerse mass, which may come from lack of observation techniques. The weight ratios of liquid droplets, biological cells, and stars will be equal to those of parcels.

Acknowledgement
The authors thank the Yukawa Institute for Theoretical Physics at Kyoto University. Discussions during the YITP workshop YITP-W-10-12 on "International and Interdisciplinary Workshop on Novel Phenomena in Integrated Complex Sciences: from Non-living to Living Systems" were useful to complete this work.

References

[1] EL-WAKIL M M 1971 Nuclear Heat Transport, Doubun Shoin.
[2] Naitoh K 1999 Oil & Gas Science and Technology **54** 205.
[3] Naitoh K 2001 Japan Journal of Industrial and Applied Mathematics 18-1 75.
[4] Naitoh K 2008 Artificial Life and Robotics **13** 10-17.
[5] Naitoh K 2010 Artificial Life and Robotics 15 117.
[6] Henley E M and Garcia A 2007 Subatomic physics. World Scientific.
[7] Naitoh K, Hashimoto K, and Inoue 2010 Proceedings of 6th WCB, Singapore.
[8] Naitoh K 2011 Proc. of the Conf. in honor of Murray Gell-Mann's 80th Birthday. World Scientific Pub.
[9] Prigogine I and Kondepudi D 1999 Thermodynamique, Editions Odile Jacob.
[10] Lowrey P and Takahashi J 2004 Annual Review of Genomics and Human Genetics **5** 407.
[11] Naitoh K 2008 Proceedings of 13th Int. Conf. on Biomedical Engineering, Springer-Verlag.
[12] Naitoh K 2009 Proceedings of the 2009 Conference on Chemical, Biological, and Environmental Engineering (CBEE), Singapore.
[13] Naitoh K 2011 Japan Journal of Industrial and Applied Mathematics, 28, 15.
[14] Naitoh K 2010 J. of Cosmology 5 999.
[15] Lamb H 1945 Hydrodynamics, 6th ed. New York: Dover.
[16] Nakamura T et al. 2009 Nucleus. Phys. Rev. Lett. 103 262501 .
[17] Takahashi K and Yamanaka S 2006 Cell **126** 663.
[18] Schumpeter J A 1939 Business cycles: A theoretical, historical, and statistical analysis of the capitalist process, (McGraw-Hill, New York).
[19] Kondratieff N 1984 The long wave cycle, (Richardson and Snyder).
[20] Bear M F, Connors B W, and Paradiso M A 2007 Neuroscience, Lippincott Williams & Wiklins Inc. USE.
[21] Naitoh K 1998 Japan Journal of Industrial and Applied Mathematics 15-1 87.
[22] Andronow A A and Chaikin C E 1949 Theory of Oscillations. Princeton University Press.
[23] Tatsumi T 1982 Fluid dynamics, Baifukan, Tokyo.
[24] Naitoh K 2006 Gene Engine and Machine Engine, Springer-Japan, Tokyo.
[25] Naitoh K 2008 Artificial Life and Robotics **13** 18-21.

Journal of Physics: Conference Series **344** (2012) 012009 doi:10.1088/1742-6596/344/1/012009

Topological aspects of the quantum spin nanotube

Yuichi Otsuka[1] **and Toru Sakai**[2]

[1]Condensed Matter Theory Laboratory, RIKEN, Wako, Saitama 351-0198, Japan

[2]JAEA SPring-8, Kouto, Sayo, Hyogo 679-5148, Japan

E-mail: sakai@spring8.or.jp

Abstract. The spin nanotube has attracted a lot of interest as an intersection between the strongly correlated electronics and the nanoscience. Among spin nanotubes we focus on the S=1/2 three-leg spin tube, because it has the largest quantum fluctuation and the strongest frustration. Using the numerical exact diagonalization of the finite-cluster Heisenberg model, we reveal that two singlet ground states are degenerated in the spin gap phase of the spin tube. It is also found that the spin gap is characterized by the quantized Berry phase.

1. Introduction

The spin nanotube[1] is one of interesting nanomaterials which are expected to be developed towards some functional devices in the near future, like the carbon nanotube. Since the S=1/2 three-leg spin tube $[(CuCl_2tach \cdot H)_3Cl]Cl_2$ was synthesized[2], several theoretical works about this systems have been published[3-11]. In the previous theoretical work by the numerical exact diagonalization and the density matrix renormalization group (DMRG) calculation [8], the spin gap was revealed to be very fragile against the lattice distortion from the regular triangle to the isosceles one, and the Berezinskii-Kosterlitz-Thouless (BKT) quantum phase transition was predicted from the gapped to the gapless phases with respect to such an asymmetry of the rung exchange interactions. A purpose of the present study is to propose a physical picture of the spin gap formation mechanism for the spin tube. Using the numerical exact diagonalization of the S=1/2 three-leg spin tube, we calculate the singlet excitation gap and the quantized Berry phase, which proposed by Hatsugai[12], to investigate topological aspects of the spin gap phase and the quantum phase transition between the gapped and the gapless Tomonaga-Luttinger liquid phases.

2. Model

We consider the S=1/2 asymmetric three-leg spin tube, shown in Fig. 1, described by the Hamiltonian

$$\hat{H} = J_1 \sum_{i=1}^{3} \sum_{j=1}^{L} \vec{S}_{i,j} \cdot \vec{S}_{i,j+1} + J_r \sum_{i=1}^{2} \sum_{j=1}^{L} \vec{S}_{i,j} \cdot \vec{S}_{i+1,j} + J'_r \sum_{j=1}^{L} \vec{S}_{3,j} \cdot \vec{S}_{1,j} \qquad (1)$$

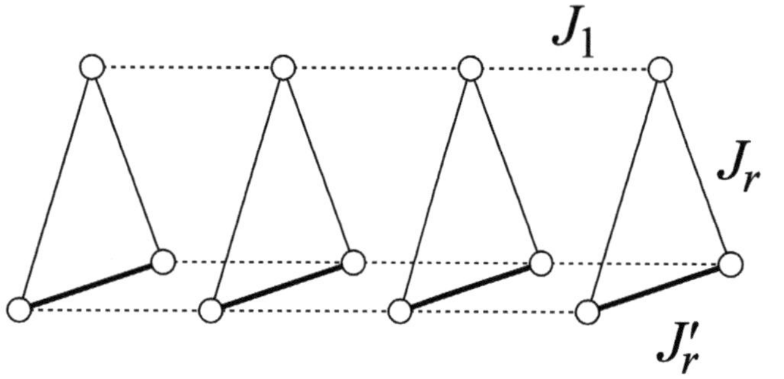

Fig. 1 Structure of the three-leg asymmetric spin tube (1).

where $\vec{S}_{i,j}$ is the spin-1/2 operator and L is the length of the tube in the leg direction. The exchange interaction constant $J_1$is for the neighbouring spin pairs along the legs, while J_r and J'_r are the rung interaction constants. All the exchange interactions are supposed to be antiferromagnetic (namely, positive).

The ratio $\alpha=J'_r/J_r$ stands for the degree of the asymmetry of the rung interactions. We will vary α and J_1 to investigate the quantum phase transitions. Throughout this paper, we fix J_r to one.

The present model includes three typical models as limiting cases; (a) α=0: the three-leg spin ladder, (b) α=1: the symmetric spin tube, and (c) $\alpha\rightarrow\infty$: the single chain plus rung dimers. Since the system is gapless in the cases (a) and (c), while gap-full in the case (b), at least two quantum phase transitions should occur with increasing α from 0 to infinity. As we already mentioned, the one-site translational symmetry along the leg ($\vec{S}_{i,j} \rightarrow \vec{S}_{i,j+1}$)is spontaneously broken in the symmetric spin tube at least in the strong-rung-coupling regime

3. Spin gap

According to the Lieb-Schultz-Mattis theorem, the spin gap phase of the S=1/2 three-leg spin tube should have two-fold degeneracy in the ground state, due to the translational symmetry break down. A schematic picture of the gapped phase for the symmetric spin tube was proposed as shown in Fig. 2, where the three equivalent singlet dimer covering patterns are resonating at each two unit cell triangles.

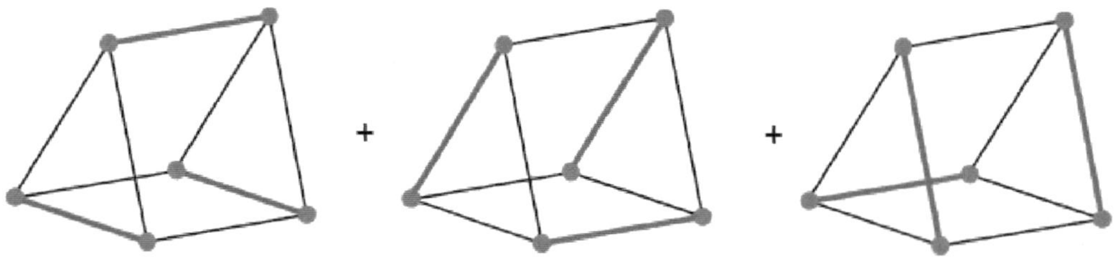

Fig. 2. A schematic picture of the spin gap formation mechanism of the S=1/2 three-leg spin tube. Circles and thick lines are spins and singlet bonds, respectively. The three equivalent patterns at each two unit cells are resonating.

If this picture is valid, a double periodicity in the chain direction should be realised in the ground state. In order to confirm it, we calculate the singlet excitation energy with the momentum k=π and test the degeneracy of it with the ground state in the thermodynamic limit. The phenomenological renormalization is a good method to determine the parameter region where the two states are degenerated in the infinite length limit. For each value of J_1, the critical point α_c is determined by the fixed point equation

$$\Delta_s(\alpha_c, N_1) = \Delta_s(\alpha_c, N_2),$$

for two different system sizes N_1 and N_2, where Δ_s is the energy difference between the singlet excitation and ground states. The calculated Δ_s is plotted versus α for L=4, 6 and 8 with fixed J1=0.5 in Fig. 3. Extrapolating the size-dependent fixed point α_c (4,6) and α_c (6,8) with respect to the system size L (we assume the size correction is proportional to $1/L^2$.), we determine α_c in the thermodynamic limit. The result is shown as solid circles in the α-J_1 plane in Fig. 4. The phase boundary between the spin gap and gapless phases determined by the phenomenological renormalization applied for the singlet-triplet gap in the previous work[8] is also plotted as a solid line in Fig. 4. Although some discrepancy between the two boundaries due to finite size effects at larger J_1, Fig. 4 suggests that the two singlet states are degenerated in the thermodynamic limit in the spin gap phase. It is consistent with that the translational symmetry is broken and double periodicity is realized in the gapped phase.

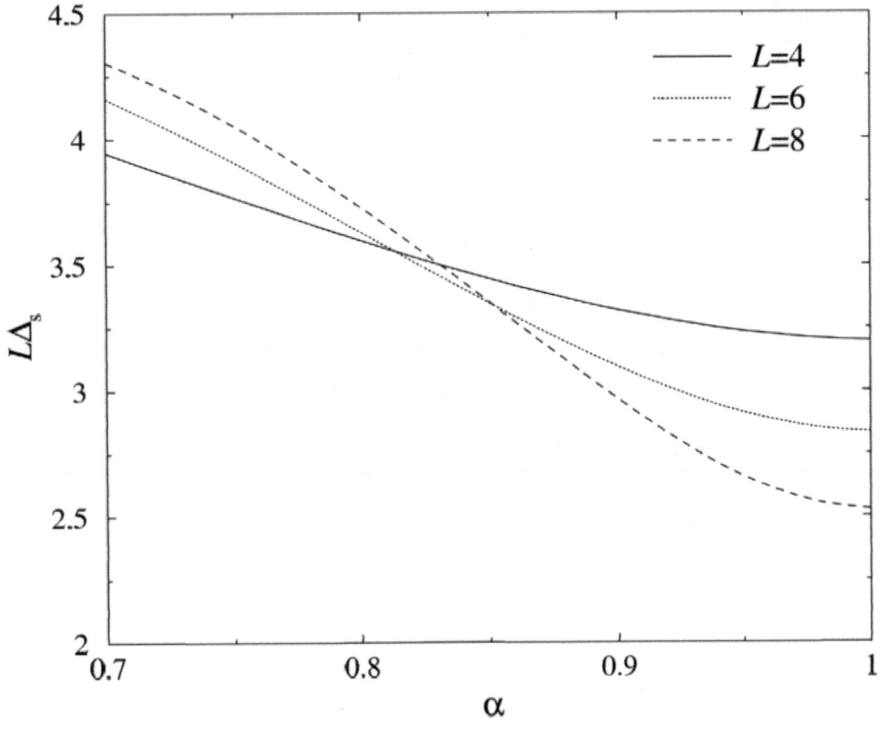

Fig. 3 Scaled singlet-singlet excitation gap $L\Delta_s$ plotted versus α with fixed J1=0.5.

Journal of Physics: Conference Series **344** (2012) 012009 doi:10.1088/1742-6596/344/1/012009

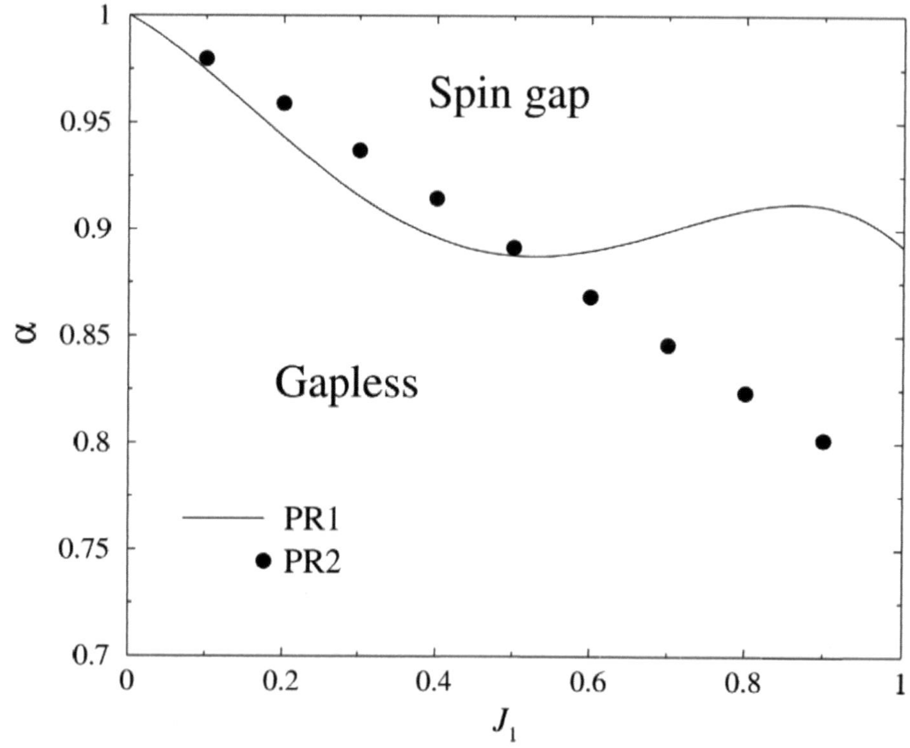

Fig. 4 Phase boundaries between the spin gap and gapless phases. Solid circles are determined by the phenomenological renormalization of the singlet-singlet excitation gap, a solid line is determined by the singlet-triplet excitation gap from Ref. [8].

4. Quantized Berry phase

The quantized Berry phase, which was proposed by Hatsugai, is a useful quantity to characterize a topological order like the string order in the Haldane gap system. It can be used to detect some hidden topological order characterized by a local gauge connection. Now we make a local SU(2) twist θ only at one link as,

$$J_{ij}\vec{S}_j \cdot \vec{S}_j \rightarrow J_{ij}/2(e^{-i\theta}S_i^+S_j^- + e^{i\theta}S_i^-S_j^+ + 2S_i^zS_j^z)$$

and sum up the Berry connection

$$\gamma = \mathrm{Arg}\prod_{C}\left\langle \psi_{\theta_j}^U \middle| \psi_{\theta_{j+1}}^U \right\rangle, \qquad |\psi_{\theta_j}^U\rangle = |\psi_{\theta_j}\rangle\left\langle \psi_{\theta_j} \middle| \phi \right\rangle$$

where $|\phi\rangle$ is a reference state to fix the gauge. If the singlet dimer covering in Fig. 2 is realised, the Berry connection is expected to quantized to be one for $\theta = \pi$ at each bond, like the string order parameter in the Haldane gap phase of the S=1 antiferromagnetic chain. The calculated γ by the numerical exact diagonalization for L=4 is plotted in the α -J_1 plane in Fig. 5. It indicates that the

Berry connection is quantized and it corresponds to one in the spin gap phase, while zero in the gapless phase. It suggests that the spin gap phase of the S=1/2 three-leg spin tube is characterized by the quantized Berry phase and it can be formulated as some topological order parameters.

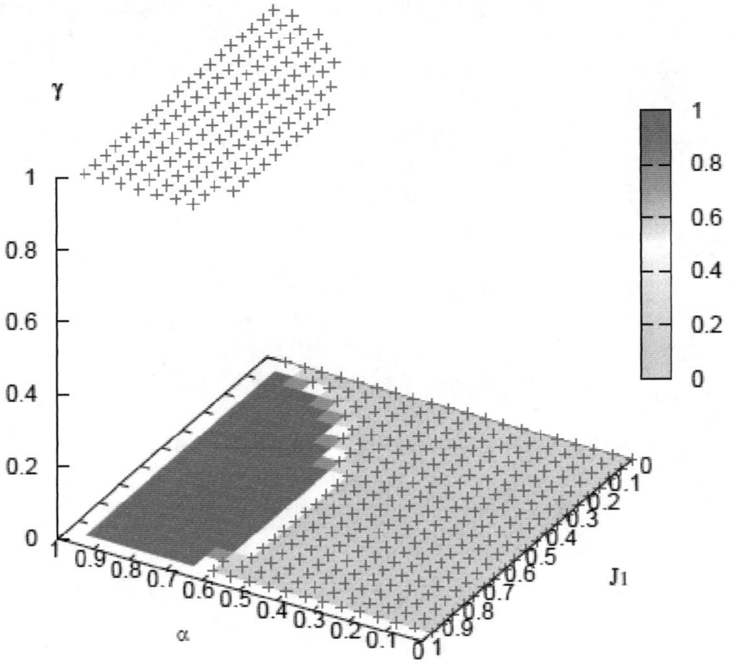

Fig. 5 Quantized Berry connection calculated by the numerical exact diagonalization for L=4.

5. Summary

The S=1/2 three-leg spin tube is investigated by the numerical exact diagonalization. It is found that the ground state has a two-fold degeneracy in the S=0 space, which is consistent with the translational symmetry break down. The spin gap phase is revealed to be characterized by the quantized Berry phase, which is related to some topological orders.

Acknowledgments

This work has been partly supported by Grants-in-Aid for Scientific Research (B) (No.17340100, No.20340096) and Priority Areas ``Invention of Anomalous Quantum Materials --- New Physics through Innovation Materials ---" (No.19014019), ``Physics of New Quantum Phases in Super-clean Materials" (No.17071011, No.18043023, No.20029020) ,``High Field Spin Science in 100T" (No.20030008, No.2003003) and" Novel states of matter induced by frustration" (No. 22014012) from the Ministry of Education, Culture, Sports, Science and Technology of Japan. We further thank the Supercomputer Center, Institute for Solid State Physics, University of Tokyo, the Cyberscience Center, Tohoku University, and the Computer Room, Yukawa Institute for Theoretical Physics, Kyoto University for computational facilities. This workshop was supported in part by the Grant-in-Aid for the Global COE Program "The Next Generation of Physics, Spun from Universality and Emergence" from the Ministry of Education, Culture, Sports, Science and Technology (MEXT) of Japan. The authors also thank the Yukawa Institute for Theoretical Physics at Kyoto University, where this work was initiated completed during the YITP-W-10-12 on "International and Interdisciplinary Workshop on Novel Phenomena in Integrated Complex Sciences: from Non-living to Living Systems".

References

[1] T. Sakai, M. Sato, K. Okunishi, K. Okamoto and C. Itoi, *J. Phys.: Condens. Matter* **22**, 403201 (2010).
[2] J. Schnack, H. Nojiri, P. Koegerler, G. J. T. Cooper and L. Cronin, *Phys. Rev. B*, **70**, 174420 (2004).
[3] T. Sakai, M. Matsumoto, K. Okunishi, K. Okamoto, and M. Sato, *Physica E*, **29** 633 (2005)
[4] J. -B. Fouet, A. Laeuchli, A. Pilgram, R. M. Noak and F. Mila, *Phys. Rev. B*, **73**, 014409 (2006).
[5] K. Okunishi, S. Yoshikawa, T. Sakai and S. Miyashita, *Prog. Theor. Phys. Suppl.*, **159**, 297 (2005).
[6] M. Sato and T. Sakai, *Phys. Rev. B*, **75**, 014411 (2007).
[7] S. Nishimoto and M. Arikawa, *Phys. Rev. B*, **78**, 054421 (2008).
[8] T. Sakai, M. Sato, K. Okunishi, Y. Otsuka, K. Okamoto and C. Itoi, *Phys. Rev. B*, **78**, 184415 (2008).
[9] D. C. Cabra, A. Honecker and P. Pujol, *Phys. Rev. B*, **58**, 6241 (1998).
[10] H. J. Schulz, in *Correlated Fermions and Transport in Mesoscopic Systems,* ed. T. Martin, G. Montambaux, J. Tran Than Van (1996); arXive:9605075.
[11] K. Kawano and M. Takahashi, *J. Phys. Soc. Jpn.*, **66**, 4001 (1997).
[12] Y. Hatsugai, *J. Phys. Soc. Jpn.*, **75**, 123601 (2006).

Synthesis and magnetic properties of Cu oxides with low-dimensional structure

M Kato, Y Nogi, S Hongo and K Hirota

Department of Molecular Chemistry and Biochemistry, Faculty of Science and Engineering, Doshisha University, Kyotanabe, Kyoto 610-0321, JAPAN

E-mail: makato@mail.doshisha.ac.jp

Abstract. We report the synthesis and magnetic properties of Cu oxides with low-dimensional structure. For $CuSb_{2-x}Ti_xO_6$ compounds with tri-rutile type structure, magnetic susceptibility χ showed that a characteristic temperature, J/k_B, where a broad maximum of χ appears due to the short-range antiferromagnetic order, decreases with increase of x. Thus, it is considered that a hole carrier doped by Ti substitution results in the decrease of J/k_B. Furthermore, it was found that the antiferromagnetic long-range order at the transition temperature (T_N), around 8.5 K for $x = 0.0$, still remains in the substituted compounds for $x = 0.3$, which indicates the one-dimensional interaction along the diagonal direction is dominant. $CuNb_{2-x}Ti_xO_6$ compounds have zig-zag chain structure in the edge-sharing CuO_6 octahedra. Temperature dependences of χ were found to be well analyzed by the alternating 1DHAF model. Hole doping into this material is realized by substituting Ti^{4+} ions for Sb^{5+} ions.

1. Introduction

Many layered Cu oxides such as high-T_c superconductors [1] generally show unconventional electric and magnetic properties based on the strongly correlated electron behaviors. Among them, tri-rutile type $CuSb_2O_6$ phase with two-dimensional (2D) tetragonal CuO_2 layers is known as a low-dimensional antiferromagnetic compound [2, 3]. Magnetic interactions between Cu atoms mainly consist of a Cu-O-Cu side direction of the square lattice (J_1) and a Cu-O-O-Cu super-exchange interaction (J_2) along the diagonal direction. These two magnetic interactions are geometrically frustrated, leading to the unconventional magnetic behaviors. Magnetic susceptibility χ of $CuSb_2O_6$ is well analyzed by the one-dimensional (1D) Heisenberg antiferromagnetic (1DHAF) model with half-integral spin [4]. On the other hand, this compound undergoes the so-called collinear ordered antiferromagnetic transition [5]. From these results, the magnetic properties of $CuSb_2O_6$ phase has been described by 1DHAF with the dominant J_2 interaction. Recently, the ground state of the 2D layered cuprates has been found to depend on the ratio of J_2/J_1; the system undergoes from Néel ordered state to a collinear state with increasing J_2/J_1. Moreover, a spin-singlet state has been predicted around the region with $J_2/J_1 \sim 0.5$ [6]. The ground state of 2D layered cuprates also can be varied by doping a carrier. A $CuNb_2O_6$ compound, with the similar chemical composition to $CuSb_2O_6$, has chains of edge-sharing CuO_6 octahedra [7], as schematically shown in Fig. 1. It also has chains of edge-sharing NbO_6 octahedra, which form washboard-like planes by the linkage of corner oxygens. The Cu atoms have the valency of +2 and carry spin $S = 1/2$, forming a quasi-1D localized spin system. For this compound, two polymorphs, orthorhombic and monoclinic, have been reported [8, 9], where the magnetic susceptibility χ of the monoclinic phase was measured in the region of temperature T between 10 K and 300 K and analyzed by introducing isolated spin dimers with the singlet-triplet excitation of 40 K. In this paper, we study synthesis and magnetic propertied of $CuM_{2-x}Ti_xO_6$ (M = Sb, Nb) compounds to investigate what magnetic characteristics are found in such low-dimensional quantum spin systems by a chemical substitution on this compound.

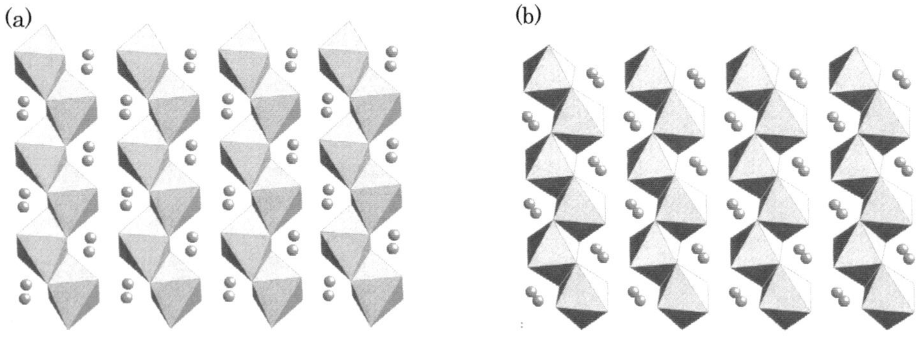

Figure 1. Schematic crystal structures of (a) orthorhombic and (b) monoclinic $CuNb_2O_6$ compounds. Blue polyhedra and red balls represent CuO_6 octahedra and Nb atoms, respectively.

2. Experimental

Samples of $CuSb_{2-x}Ti_xO_6$ ($0.0 \leq x \leq 0.6$) were prepared by a conventional solid state reaction (SSR). Starting materials of CuO (purity 99.999%), Sb_2O_3 (99.99%) and TiO_2 (99.99%) were mixed carefully, subsequently pressed in the form of a cylindrical pellet. Then, the pellet was heated in air at 750°C for 1 day. The resulting intermediate is reground, pressed and then heated further for 2 days at 800°C.

It has been difficult to synthesize $CuNb_{2-x}Ti_xO_6$ compounds with a conventional SSR method. Thus, polycrystalline samples of $CuNb_{2-x}Ti_xO_6$ phases were synthesized by citric acid complex polymerization (CCP) methods. Starting materials of $Cu(NO_3)_2{\cdot}3H_2O$ (99%), $Nb(OC_2H_5)_5$ (99.99%) and Ti(O-i-$C_3H_7)_4$ (99.999%) were dissolved in $C_3H_4(OH)(COOH)_3$ (99%) and HOC_2H_4OH (99.5%) solution, then heated at 60-80°C for 12 hours. Obtained polymerized complex was dried at 150°C for 12 hours. The precursor was heated at 400°C for 1day to decarbonate, and then heated at 650-700°C for 2 days with intermediate grinding.

The phase identification was performed by X-ray diffraction measurements (Rigaku, Rint-2200) using the monochromatic Cu Kα radiation. Magnetic susceptibility (χ) was measured by SQUID magnetometer (Quantum Design, MPMS-5T) in the temperature range from 2 to 300 K under the magnetic field of 0.1 or 1 T. In the measurement of χ, the magnetic field was applied before cooling the sample (Zero-field cooling, ZFC) because none of the sample is ferromagnetic.

3. Results and Discussion

In Fig. 2, we show XRD patterns for $CuSb_{2-x}Ti_xO_6$ ($0.0 \leq x \leq 0.6$) samples obtained by SSR methods. Samples with Ti composition of $0.0 \leq x \leq 0.3$ were found to be indexed by monoclinic tri-rutile type structure. The sample with higher Ti content indicates small diffraction peaks due to CuO, although the lattice parameters were found to vary with increasing x. Thus, we have succeeded in synthesize the solid solution of $CuSb_{2-x}Ti_xO_6$ with tri-rutile type structure up to $x = 0.6$.

Temperature dependences of magnetic susceptibility for $CuSb_{2-x}Ti_xO_6$ ($0.0 \leq x \leq 0.6$) are shown in Fig. 3. The sample of $x = 0$ indicates the broad maximum at about 60 K due to the

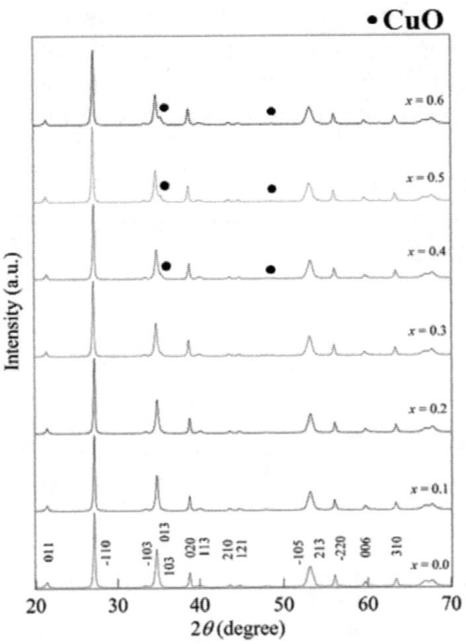

Figure 2. XRD patterns of $CuSb_{2-x}Ti_xO_6$ with tri-rutile type structure.

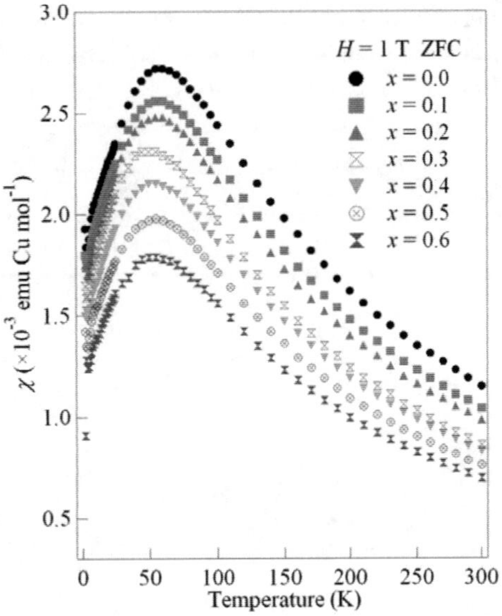

Figure 3. Temperature dependence of magnetic susceptibility for $CuSb_{2-x}Ti_xO_6$ compounds with tri-rutile type structure. The Curie-Weiss paramagnetic components are subtracted from the raw data for the samples with $x \geq 0.1$.

short-range antiferromagnetic interaction of low-dimensional structure as previously reported [3]. The Curie-Weiss paramagnetic components are subtracted from the raw data for the samples with $x \geq 0.1$ using the following estimation. The data of magnetic susceptibility were analyzed by the 1DHAF model presented by Bonner and Fisher (BF) [4] with CW term as follows.

$$\chi = (1 - A)\chi_{\rm BF} + A\chi_{\rm CW} + \chi_0 \tag{1}$$

$$\chi_{\rm BF} = \frac{N_{\rm A} g^2 \mu_{\rm B}^2}{k_{\rm B}} \times \frac{0.25 + 0.14995z + 0.30094z^2}{1 + 1.9862z + 0.68854z^2 + 6.60626z^3} \tag{2}$$

$$\chi_{\rm CW} = \frac{N_{\rm A} g'^2 \mu_{\rm B}^2 S(S+1)}{3k_{\rm B}(T - \Theta)} \tag{3}$$

Here, A represents the ratio of BF and CW term, χ_0 a constant parameter of χ, $N_{\rm A}$ Avogadro' s number, g (g') g-factor, $\mu_{\rm B}$ Bohr magnetron, $k_{\rm B}$ Boltzmann constant, S spin quantum number, Θ Weiss temperature, and z denotes $J/k_{\rm B}T$. The raw data of temperature dependence of χ were performed by least-square curve fitting with parameters of J, g (g'), A and χ_0. As a result, the value of $J/k_{\rm B}$ decreases with increasing x. We summarize the values of $J/k_{\rm B}$, and $T_{\rm N}$ in Fig. 4 with data for other solid solutions of $CuSb_{2-x}T_xO_6$ (T = Sn, Ta, Nb) [10]. One can note that $J/k_{\rm B}$ for Ta and Nb-substituted system has still finite value at higher concentration. This can be caused by the invariant carrier concentration because Ta and Nb is pentavalent as same as Sb. On the other hand, $J/k_{\rm B}$ for Sn-substituted system rapidly decrease with x due to the doped holes by tetravalent Sn. Thus, the rapid decrease of $J/k_{\rm B}$ is also expected for Ti-substituted system, however, the experimental data of $J/k_{\rm B}$ are almost intermediate between

Ta- and Nb-substituted systems. In the present stage, the reason of the magnetic behavior of Ti-system is unclear yet, although these behaviors should be caused by structural and magnetic interactions between Cu atoms in 1DHAF. Furthermore, it was found that the antiferromagnetic long-range order at the transition temperature (T_N), around 8.5 K for $x = 0.0$, still remains in the substituted compounds for $x = 0.3$, which indicates the one-dimensional interaction along the diagonal direction is dominant.

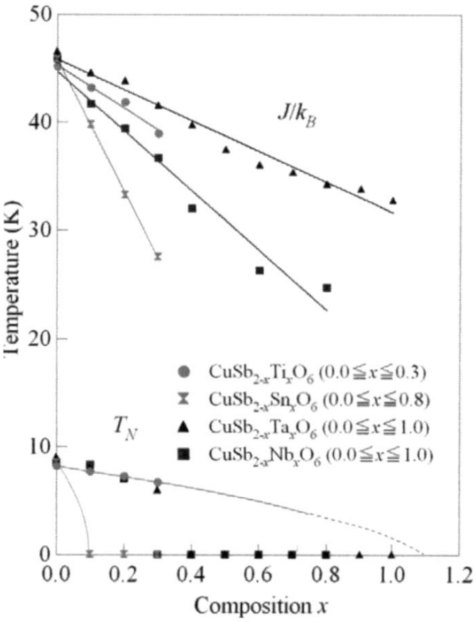

Figure 4. Composition dependences of magnetic parameters for $CuSb_{2-x}T_xO_6$ (T = Ti, Sn, Ta, Nb) estimated from the analysis using BF model.

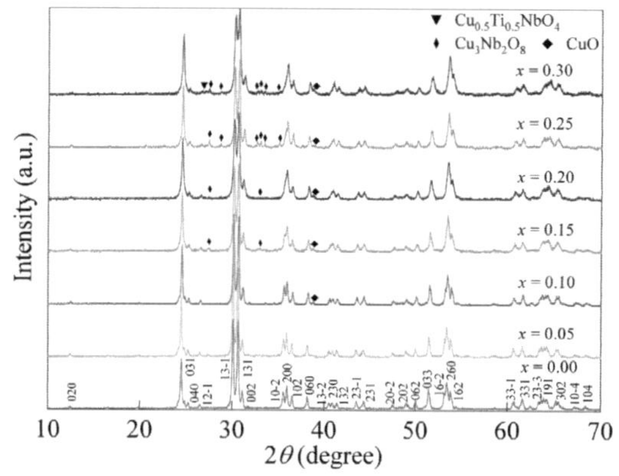

Figure 5. XRD patterns for $CuNb_{2-x}Ti_xO_6$ compounds obtained by CCP methods.

We show typical XRD patterns for $CuNb_{2-x}Ti_xO_6$ compounds obtained by CCP methods in Fig. 5. Samples with Ti composition of $0.0 \le x \le 0.1$ were found to be indexed by monoclinic $CuNb_2O_6$ (m-$CuNb_2O_6$) type structure. The m-$CuNb_2O_6$ type structure has been commonly synthesized by SSR method at low temperatures of about 700°C, although Ti-substituted system has not been obtained. On the other hand, single phase of orthorhombic $CuNb_2O_6$ (o-$CuNb_2O_6$) compound, which is higher temperature phase, could not be synthesized by SSR nor CCP method. The sample with higher Ti content indicates small diffraction peaks due to impurity phases, although the lattice parameters were found to vary with increasing x. Thus, we have succeeded in synthesize the solid solution of m-$CuNb_{2-x}Ti_xO_6$ compounds up to $x = 0.3$.

Temperature dependences of magnetic susceptibility for $CuNb_{2-x}Ti_xO_6$ ($0.0 \le x \le 0.3$) are shown in Fig. 6. The sample of $x = 0$ indicates the broad maximum at about 30 K due to the spin-gap formed by spin dimers on the zig-zag 1DHAF chains. The peaks of χ shift towards low temperatures and almost disappears for x =0.3, which means that the magnetic interaction between neighbor atoms decreases with x. These data were found to be successfully analyzed by the alternating 1DHAF model proposed by Johnston (Johnston model) as follows [11].

$$\chi = (1 - A)\chi_{AE} + A\chi_{CW} + \chi_0 \quad (4)$$

In Eq. (4), χ_{AE} represents the magnetic susceptibility of the alternating 1DHAF chains with

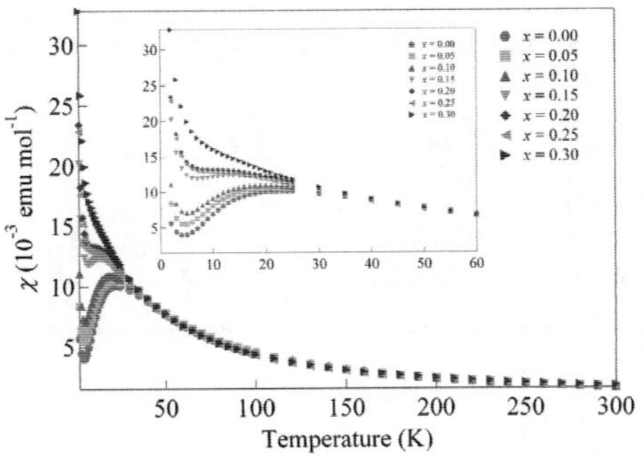

Figure 6. Temperature dependence of magnetic susceptibility for m-$CuNb_{2-x}Ti_xO_6$ compounds.

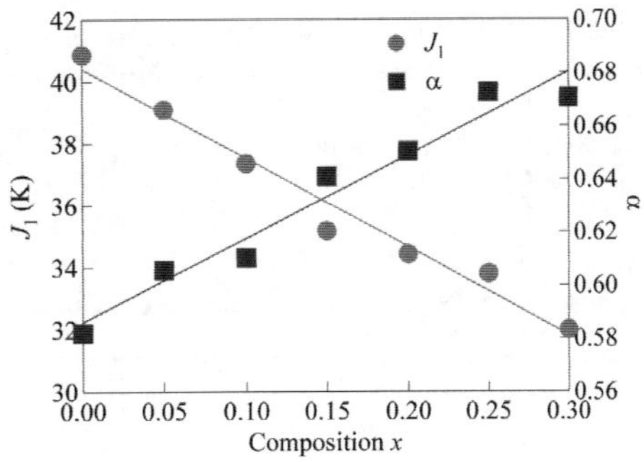

Figure 7. Composition dependences of magnetic parameters for m-$CuNb_{2-x}Ti_xO_6$ estimated from the analysis using Johnston model.

a parameter of α, the ratio of J_2/J_1; the system is a uniform 1DHAF chain (BF model) for $\alpha = 1$ and an isolated dimer model for $\alpha = 0$. Numerical formula should be referred to Ref. [11]. As a result, we have obtained magnetic parameters of α and J_1 as shown in Fig. 7. One can note that J_1 decreases with x as mentioned above, and that α increases with carrier doping. The increase in α indicates that the system changes from alternating to uniform 1DHAF. Moreover, the doped holes are considered to be localized on 1DHAF chains because the temperature dependence of χ is almost invariant at higher temperatures as shown in Fig. 6. From these behaviors, it should be concluded that the doped hole carriers result not in the itinerancy of the system but in the enhancement of magnetic exchange interaction on 1DHAF chains. Since these behaviors are similar to those of high-T_c Cu oxides, it is necessary to investigate the microscopic properties of strongly correlated electrons in $CuNb_{2-x}Ti_xO_6$ compounds, which will be reported elsewhere.

4. Conclusion

The samples of solid-solution compounds, $CuSb_{2-x}Ti_xO_6$ and $CuNb_{2-x}Ti_xO_6$ were successfully synthesized using SSR and CCP methods, respectively. From magnetic susceptibility measurements for $CuSb_{2-x}Ti_xO_6$ compounds, it is shown that a characteristic temperature, J/k_B, where a broad maximum of χ appears due to the short-range antiferromagnetic order, decreases with increase of x. Thus, it is considered that a hole carrier doped by Ti substitution results in the decrease of J/k_B. Hole doping into $CuNb_2O_6$ is realized by substituting Ti^{4+} ions for Sb^{5+} ions. The doped hole carriers result not in the itinerancy of the system but in the enhancement of magnetic exchange interaction on 1DHAF chains.

Acknowledgments

This work was supported by "Advanced Study for Integrated Particle Science and Technology," Strategic Development of Research Infrastructure for Private Universities, #S0901039, the Ministry of Education, Culture, Sports, Science and Technology (MEXT), Japan and also supported by a Grant-in-Aid Scientific Research of Japan Society for Promotion of Science (22550129).

References

[1] Bednorz J G and Müller K A 1986 *Z. Physik.* B **64** 189
[2] Shimada S, Kodaira K and Matsushita T 1985 *J. Cryst. Growth* **72** 753
[3] Nakua A, Yun H, Relmers J N, Greedan J E and Stager C V 1991 *J. Solid State Chem.* **91** 105
[4] Bonner J C and Fisher M E 1964 *Phys. Rev.* **135** A640

[5] Kato M, Hatazaki A, Yoshimura K and Kosuge K 2000 *Physica* B **281-282** 663
[6] Melzi R, Carretta P, Lascialfari A, Mambrini M, Troyer M, Millet P and Mila F 2000 *Phys. Rev. Lett.* **85** 1318
[7] Drew M G B, Hobson R J and Padayatcy V T 1993 *J. Mater. Chem.* **3** 889
[8] Wahlström E and Marinder B-O 1977 *Inorg. Nucl. Chem. Lett.* **13** 559
[9] Drew M G B, Hobson R J and Padayatcy V T 1995 *J. Mater. Chem.* **5** 1779
[10] Kato M, Ishii T, Kajimoto K, Yoshimura K and Kosuge K 2002 *J. Phys. Chem. Solids* **63** 1129
[11] Johnston D C, Kremer R K, Troyer M, Wang X, Klümper A, Bud'ko S L, Panchula A F and Canfield P C 2000 *Phys. Rev.* B **61** 9558

Journal of Physics: Conference Series **344** (2012) 012011 doi:10.1088/1742-6596/344/1/012011

Geometric frustration in the stella quadrangula lattice and metamagnetism in Fe_3Mo_3N

H. Nakamura, S. Terazawa, T. Waki, Y. Tabata, M. Takigawa[†]

Department of Materials Science and Engineering, Kyoto University, Kyoto 606-8501, Japan
[†] Institute for Solid State Physics, University of Tokyo, Kashiwa 277-8581, Japan

E-mail: nakamura.hiroyuki.2w@kyoto-u.ac.jp

Abstract. η-carbide-type transition-metal compounds, which include the *stella quadrangula* lattice, are good candidates for testing the geometric frustration in the itinerant electron magnet. The broad peak observed in the temperature dependence of the susceptibility for one of the η-carbide-type bimetallic nitrides, Fe_3Mo_3N, has been confirmed microscopically by ^{95}Mo NMR Knight-shift measurements. The itinerant electron metamagnetic transition observed in this compound is discussed in connection with the $\chi(T)$ behavior.

1. Introduction

Transition-metal nitrides and carbides with the η-carbide-type structure are expected to show new and various electronic phenomena and functions, because they are typical intermetallic compounds with narrow d-electron bands. However, despite that a huge number of the compounds have been known to exist, their electronic properties have been less studied in contrast to extensive investigations in material science, for example, as hard and refractory materials, and as potential catalysts [1]. Recently, we have studied systematically electronic properties of the η-carbide-type compounds, and already reported several examples of interesting discoveries [2, 3, 4]. In this article, we briefly summarize geometrical characteristics of the η-carbide-type structure, emphasizing possible geometric frustration of the magnetic interaction in the *stella quadrangula* lattice found in this structure [5]. As an example of the experimental data, we show the results of NMR measured for a η-carbide-type nitride Fe_3Mo_3N, which shows quantum criticality and sharp itinerant-electron metamagnetism.

2. Geometrical characteristics of the η-carbide-type structure

The prototypes of η-carbide-type compounds are Fe_3W_3C and Ni_6Mo_6C. For both cases, the space group is cubic $Fd\bar{3}m$ (227). 3d elements (Fe and Ni) occupy two crystallographic sites with the Wyckoff notations 16*d* and 32*e*. 4d and 5d elements (Mo and W) occupy the 48*f* site. The difference between Fe_3W_3C- and Ni_6Mo_6C-type structures is the atomic position of C. In the former, C atoms occupy the 16*c* site but in the latter the 8*a* site. A number of carbides and nitrides with the same structure are known to exist. The general chemical formulas can be written as $(T_2T')M_3X$ and $(T_4T'_2)M_6X$. When T and T′ sites are occupied by the same element as Fe in Fe_3W_3C, the sublattice composed of the 16*d* and 32*e* sites is called the *stella quadrangula* lattice [5], which plays an essential role in determining magnetic properties of the η-carbide-type bimetallic compounds. When a 3d element occupies the metallic sites, the narrow

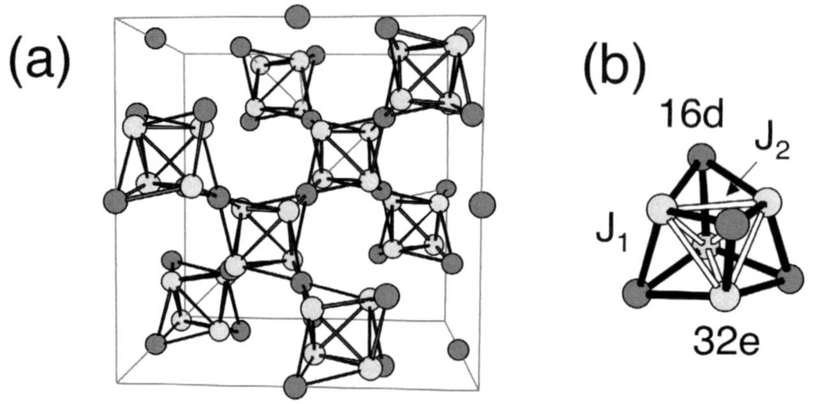

Figure 1. (a) The *stella quadrangula* lattice (Wyckoff positions $16d$ and $32e$) found in the η-carbide-type structures (space group $Fd\bar{3}m$, in this figure, the origin choice 2 was selected). (b) The *stella quadrangula* as an element of the *stella quadrangula* lattice. J_1 (black bonds) and J_2 (white bonds) correspond to NN ($16d$–$32e$) and NNN ($32e$–$32e$) interactions, respectively.

d bands are likely to sit at around the Fermi level [6], and dominate the magnetic properties. As a result, the η-carbide-type compounds behave as typical itinerant electron magnets with metallic conductivity. On the other hand, C and N atoms, with their p states far below the Fermi level, play an important role in their mechanical properties.

The *stella quadrangula* lattice is schematically shown in figure 1(a). The element of this lattice is a '*stella quadrangula*' (figure 1(b)), which means a stellate (star-shaped) tetrahedron, or a tetra-capped tetrahedron. In a different viewpoint, the *stella quadrangula* can be regarded as two nested regular tetrahedrons having the same center of gravity. The *stella quadrangula* lattice is a corner-shared lattice of *stella quadrangulae* just like the pyrochlore lattice as the corner-shared tetrahedral lattice. In the *stella quadrangula*, there are two types of near-neighbor bonds, $16d$–$32e$ and $32e$–$32e$. These nearest-neighbor (NN) and next-nearest-neighbor (NNN) distances are nearly the same, for which a direct exchange coupling is expected, while third neighbors ($16d$–$16d$) are much apart. In fact NN and NNN distances depend on the atomic coordinate of $32e$ (zzz), where z should be determined experimentally. In particular, when $z = 0.3$, the NN and NNN distances are the exactly same, and experimental values close to $z = 0.3$ have been reported for actual η-carbide-type compounds. Thus it is reasonable to consider only J_1 and J_2 corresponding to the NN and NNN interactions, respectively, and neglect those of further neighbors.

To visualize and speculate the general nature of the *stella quadrangula* lattice, it is useful to examine the two dimensional version of the *stella quadrangula*, which is a tri-capped triangle as shown in figure 2. Here we take account of only the outer edge interaction J_1 and inner triangle interaction J_2, and consider the arrangement of Ising spins. As in figure 2(a), when J_1 is negative and J_2 is zero, the elements is not frustrated (open and closed circles represent up and down spins). However, once negative J_2 comparable to J_1 is introduced, the element turns to be a typical frustrated lattice (figure 2(b)). As another possibility, when J_1 is positive and J_2 is zero, needless to say, the element is not frustrated, but the same situation occurs when negative J_2 comparable to J_1 is introduced. These facts leads to the following speculations for the *stella quadrangula*: (i) The element is frustrated only when J_2 is negative and dominant. (ii) J_1 works to release the frustration. (iii) The frustration does not depend on the sign of J_1. In other words, the frustrated and non-frustrated states are switched depending on the

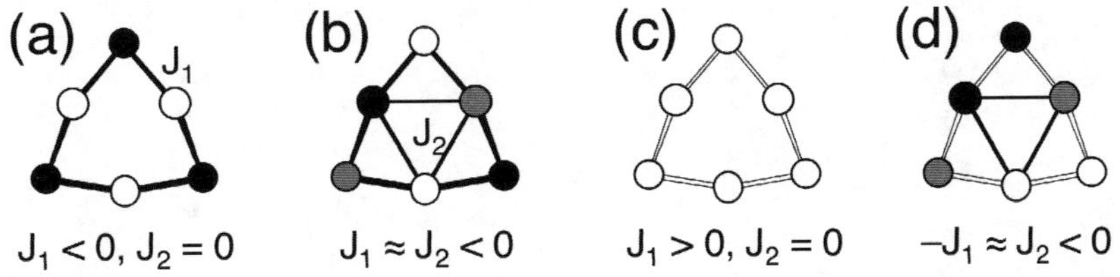

Figure 2. Ising spins in a tri-capped triangle, two-dimensional version of the *stella quadrangula*. We denote outer-edge and inner-triangle interactions as J_1 and J_2, respectively. Open and closed bonds represent ferromagnetic and antiferromagnetic interactions, respectively. Open and closed circles imply up and down spins, and 'gray' circles unfixed spins. The cases (a) and (c) with $J_2 = 0$ are not frustrated, while (b) and (d) with negative J_2 comparable to J_1 are frustrated.

ratio of $J_2/|J_1|$. The presence of the same general trends were verified for Heisenberg spins in the infinite corner-shared network of *stella quadrangulae*, by calculating the dispersion of the Fourier transform of the exchange integral matrix for the unit cell of the *stella quadrangula* lattice with assuming various J_2/J_1 ratios [3]. In conclusion, in the η-carbide-type bimetallic magnets, we expect subtle transitions between frustrated and non-frustrated states associated with the variation in the $J_2/|J_1|$ ratio.

3. Susceptibility peak and metamagnetism in Fe_3Mo_3N

One of the most interesting η-carbide-type compounds is Fe_3Mo_3N, which shows a quantum critical behavior at low temperature [2] and a quite sharp itinerant-electron metamagnetic transition at ~14 T [4]. The exchange-enhanced Pauli paramagnetism of Fe_3Mo_3N is characterized by a broad peak in the temperature dependence of the susceptibility, observed at ~80 K [2]. Here, as a microscopic evidence of the $\chi(T)$ peak, we show the result of the Knight shift estimated from the ^{95}Mo NMR. A spin-echo NMR experiment was done with a phase-coherent type spectrometer at ISSP, the University of Tokyo. Figure 3 shows an example of Mo NMR spectrum measured at 300 K and under the external field of 10.51 T. The Mo element has two observable isotopes; ^{95}Mo (natural abundance 15.9%) with the nuclear spin $I = 5/2$, the gyromagnetic ratio $\gamma = 2.774$ MHz/T and the quadrupole moment $Q = -0.12 \times 10^{-24}$ cm^2, and ^{97}Mo (9.5%) with $I = 5/2$, $\gamma = 2.832$ MHz/T and $Q = -1.1 \times 10^{-24}$ cm^2. Since the site symmetry of the Mo site ($48f$) is $2mm$, the asymmetric quadrupole interaction is expected at the Mo site. As seen in figure 3, the spectrum looks with the quadrupolar satellites, but the ^{95}Mo and ^{97}Mo components are not resolved due to their close γ values. Hence we take the peak position in the spectrum as the resonance frequency of the ^{95}Mo component, and estimate the ^{95}Mo Knight shift at various temperature between 4.2 and 300 K. The temperature dependence of the ^{95}Mo Knight shift thus obtained is plotted in figure 4(a) together with the data of macroscopic susceptibility. At high temperatures, the Knight shift exactly follows the temperature dependence of the susceptibility and takes a maximum at ~80 K. Thus the presence of the broad $\chi(T)$ peak is evidenced form a microscopic viewpoint. At low temperatures, the Knight shift approaches a constant value, in contrast to the critical divergence of the susceptibility. Here it should be noted that the susceptibility was measured under a relatively low field of 1 T, while the Knight shift at ~10 T. The low-temperature $\chi(T)$ upturn is ascribed to either the essential quantum critical divergence associated with the non-Fermi liquid behavior or the Curie term of the paramagnetic impurities. In the former case, since the quantum critical behavior is suppressed by the high magnetic field of the order of 10 T, as evidenced in specific heat measurements [2], no divergence in the Knight shift measured under the high field is reasonable. In figure 4(b), the Knight shift is plotted against the susceptibility with temperature as an

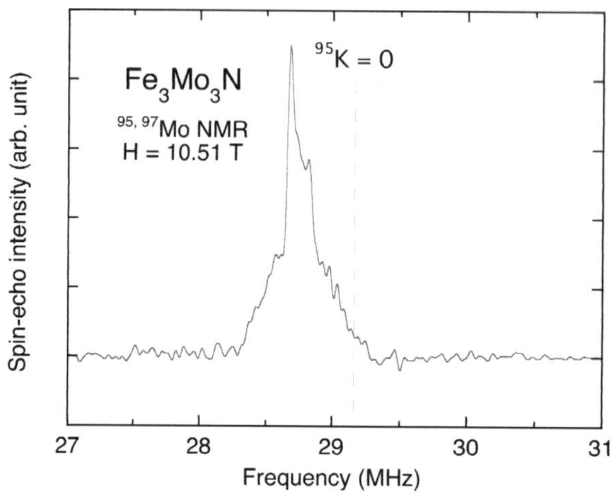

Figure 3. An example of Mo NMR spectrum measured for Fe_3Mo_3N at 300 K and under the external field of 10.51 T. The broken line represents the zero Knight shift position of ^{95}Mo.

implicit parameter. The high-temperature linear part to the K–χ plot gives an approximate value of the hyperfine coupling -6.0 T/μ_B, which is compared with the typical on-site hyperfine coupling constant expected for the Mo nucleus ~ -40 T/μ_B, suggesting that the contribution of the Mo magnetism is rather minor but the Fe sublattice forming the *stella quadrangula* lattice plays an principal role in the magnetism of Fe_3Mo_3N.

In general, the $\chi(T)$ peak is closely related to the occurrence of the itinerant electron metamagnetism. Yamada developed a simple theory based on the free energy expansion of the magnetizaiton M by taking account of the effect of spin fluctuations, and succeeded in explaining itinerant electron metamagnetic transitions of various materials [7]. In the classical theory, the itinerant electron metamagnetism takes place under the conditions $a > 0$, $b < 0$, and $c > 0$ with $3/16 < ac/b^2 < 9/20$, where a, b, and c are expansion coefficients given in the free-energy expansion up to the M^6 term as

$$F = F_0 + \frac{1}{2}aM^2 + \frac{1}{4}bM^4 + \frac{1}{6}cM^6. \tag{1}$$

Yamada proposed a relation for estimating ac/b^2 from $\chi(T)$ as

$$\frac{ac}{b^2} = \frac{5}{28}\left[1 - \frac{\chi(0)}{\chi(T_{\max})}\right]^{-1}, \tag{2}$$

where $T_{\max}$ is the temperature where $\chi(T)$ takes a maximum. We apply this relation to Fe_3Mo_3N. By extrapolating the high-temperature Knight shift data to the lowest temperature and excluding the effect of the low-temperature critical divergence, intrinsic $\chi(T_{\max})$ and $\chi(0)$ are estimated to be 18.2×10^{-3} and 12.7×10^{-3} emu/(f.u. mol), respectively. With these values, ac/b^2 is estimated as 0.59, which is outside the upper bound of the criterion $9/20 = 0.45$, suggesting that Yamada's theory is too simple to explain the itinerant electron metamagnetism of Fe_3Mo_3N. One of the most plausible reasons for the discordance is that the magnetic order parameter M is not uniform in this system owing to the presence of two different magnetic sites in the Fe sites. In other words, the magnetic state of this compound is determined as a result of the competition between J_1 and J_2 in the *stella quadrangula* lattice as mentioned above. In this case, the metamagnetic transition is interpreted as the field-assisted release of the frustration from

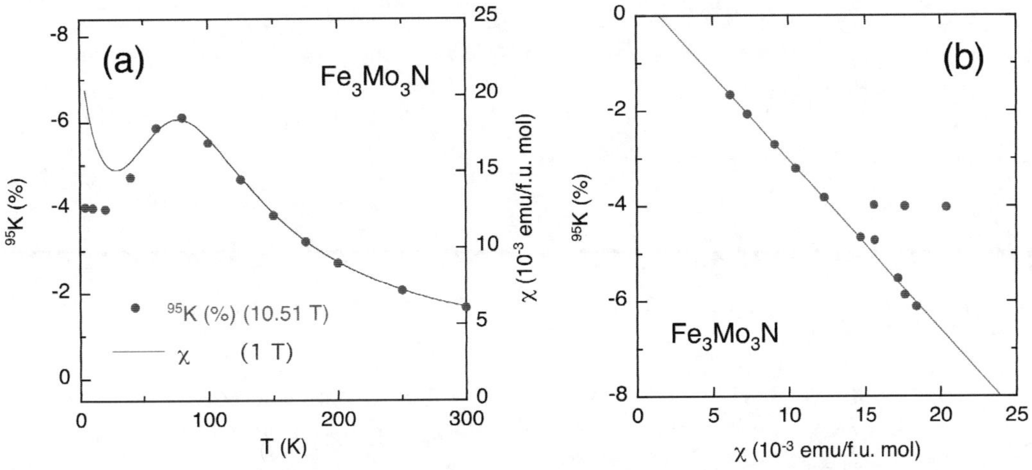

Figure 4. (a) The temperature dependence of the ^{95}Mo Knight shift measured for Fe_3Mo_3N. The susceptibility (solid curve) is also shown. The Knight shift and the susceptibility were measured under the fields of 10.51 and 1 T, respectively. (b) The K–χ plot for ^{95}Mo NMR. The linear relation at high temperatures (straight line) gives the hyperfine coupling constant $A_{\rm hf} = -6.0\ {\rm T}/\mu_{\rm B}$.

the frustration-induced non-ordered ground state. A similar discordance between experimental result and Yamada's theory has already been reported for another η-carbide-type bimetallic compound Co_3Mo_3C [3].

Acknowledgments
The authors are indebted to Prof. Y. Takahashi for theoretical support. This study was supported by a Grant-in-Aid for Scientific Research on Priority Areas "Novel States of Matter Induced by Frustration" (19052003), a Grant-in-Aid for the Global COE Program, "International Center for Integrated Research and Advanced Education in Materials Science", and a Grant- in-Aid for Young Scientists (B) 21760531 from the Ministry of Education, Culture, Sports, Science and Technology of Japan. The authors thank the Yukawa Institute for Theoretical Physics at Kyoto University. Discussions during the YITP workshop YITP-W-10-12 on "International and Interdisciplinary Workshop on Novel Phenomena in Integrated Complex Sciences: from Non-living to Living Systems" were useful to complete this work.

References
[1] for example,
Toth L E 1971 *Transition Metal Carbides and Nitrides* Academic Press, New York, 1971
Jacobsen C J H 2000 *Chem. Commun.* 1057
Errandonea D *et al.* 2010 *Phys. Rev.* B **82** 174105
[2] Waki T *et al.* 2010 *J. Phys. Soc. Jpn.* **79** 043701
[3] Waki T *et al.* 2010 *J. Phys. Soc. Jpn.* **79** 093703
[4] Waki T *et al.* 2011 *EPL* **94** 37004
[5] Nyman H, Andersson S and O'Keeffe M 1978 *J. Solid State Chem.* **26** 123
[6] for example, Suetin D V *et al.* 2009 *Physica* B **404** 3544
[7] Yamada H 1993 *Phys. Rev.* B **47** 11211

Superconducting properties and thermal expansion of $YbGa_xSi_{2-x}$

Naohito Tsujii, Motoharu Imai and Hideaki Kitazawa

National Institute for Materials Science, 1-2-1 Sengen, Tsukuba 305-0047, Japan

E-mail: tsujii.naohito@nims.go.jp

Abstract. $YbGa_xSi_{2-x}$ forms the hexagonal AlB_2-type crystal structure in the chemical composition range of $1.12 \leq x \leq 1.49$ and shows superconductivity with critical temperatures $T_C \leq 2.5$ K. In this paper, we investigate the electronic and thermal properties of the $YbGa_xSi_{2-x}$ compounds by means of electrical resistivity (ρ) and specific heat (C) measurements under magnetic fields, as well as by the temperature dependent x-ray diffraction study. The onset temperature of superconductivity, T_C^{onset}, decreases with the increase of the Ga content x, from 2.5 K for $x = 1.15$ to 1.9 K for $x = 1.31$. Estimated upper critical field at zero temperature, $H_{C2}(0)$, concomitantly decreases from 5.9 kOe for $x = 1.15$ to 3.1 kOe for $x =$ 1.31. The normal-state electronic specific-heat coefficient is found to be almost independent to x; $\gamma = 8\text{mJ/molK}^2$ for $x = 1.18$, 1.31, and 1.41. The small value of γ indicates electron-electron correlation is weak in this compound. An irreversible behavior is observed in $\rho(T)$. $\rho(T)$ of $YbGa_{1.2}Si_{0.8}$ shows a hump-like increase below about 200 K on cooling while it shows a rather monotonic increase on heating. Powder x-ray diffraction of this sample taken at 12 K shows no evidence of structural phase transition. Temperature dependence of the lattice parameters in this sample show normal behavior: both a and c decrease with decreasing temperature with thermal expansion coefficients similar to those in YbGaGe. Possible mechanism for the irreversible behavior in $\rho(T)$ is discussed.

1. Introduction

Superconductors with nearly two-dimensional crystal structures are extensively investigated because of their high T_C. A series of ternary silicides M(Ga,Al)Si (M = Ca, Sr, Ba) have been reported to crystallize in the hexagonal AlB_2-type structure and show superconductivity [1, 2, 3, 4, 5, 6, 7, 8]. Since this compound is isostructural with MgB_2, in which $T_C = 39$ K is reported [9], similarities in the electronic structures between them are expected. It has, however, been found that the ternary silicide superconductors have quite contrasting properties with those of MgB_2. For MgB_2, holes in the honeycomb layer of boron are dominant carriers [10, 11, 12], and in-plain phonon mode of the boron layer is responsible for superconductivity [13, 14]. On the other hand, dominant carriers in the ternary silicides are electrons, as suggested from the negative Seebeck coefficient [5, 7] and the Hall effect measurements [15]. Softening of the out-of-plain phonon mode is observed for CaAlSi by inelastic x-ray or neutron scattering measurements [16, 17], in contrast to the in-plain mode for MgB_2. Therefore the phonon mode responsible for superconductivity is also different between MgB_2 and the ternary silicides. This fact suggests the presence of variety of superconducting mechanisms among the AlB_2-type superconductors, pointing to the importance of further studies of the physical and structural properties in this class of compounds.

Recently, a new compound $YbGa_{1.1}Si_{0.9}$ with the AlB_2-type structure has been discovered to be superconducting below $T_C = 2.4$ K [18]. Later it has been revealed that the AlB_2-type structure is stabilized for $YbGa_xSi_{2-x}$ with $1.12 \leq x \leq 1.49$ [19]. T_C decreases linearly with x from 2.4 K for $x = 1.15$ to less than 1.8 K for $x = 1.41$. The Yb ion was considered to be in the nonmagnetic divalent state, but the precise x-ray emission spectroscopy reveals a slight mixed-valent state $Yb^{2.3+}$, which does not depend on temperature and/or chemical composition [19]. Quite recently, it has been demonstrated that T_C decreases rapidly by applying pressure. The coefficient of the decrease of T_C against pressure, dT_C/dP, is one order of magnitude greater in $YbGa_xSi_{2-x}$ than in CaGaSi [20]. This can be due to the hybridization of f electrons with conduction electrons, as inferred from the x-ray emission spectroscopy measured under high pressures [21].

In this Workshop [1], we report electrical resistivity and specific heat of $YbGa_xSi_{2-x}$ polycrystalline samples measured in magnetic fields. We observe an irreversible behavior in the temperature dependence of the electrical resistivity. We also performed the temperature dependent x-ray diffraction measurements for $YbGa_{1.2}Si_{0.8}$ powder sample to examine any structural phase transition at low temperature and also to investigate the temperature dependence of the lattice parameters.

2. Experimental

Polycrystalline samples of $YbGa_xSi_{2-x}$ were synthesized by argon arc melting and subsequent annealing, as described earlier [19]. The chemical composition was confirmed by the electron probe micro analysis (EPMA), for which single crystals of GaAs, Si, and $YbAl_3$ were used as the standard samples for chemical composition calculation. Here, the single crystal of $YbAl_3$ was grown by an Al-flux method in an alumina crucible [22, 23]. Electrical resistivity was measured by a four probe method using a Physical Property Measurement System (PPMS), Quantum Design Co. Specific heat measurement was carried out by a relaxation method using a PPMS. Powder x-ray diffraction patterns of $YbGa_{1.2}Si_{0.8}$ was taken with a Rigaku RINT2000 diffractometer using Cu $K\alpha$ radiation operating at 40 kV and 200 mA from room temperature down to 12 K. The x-ray diffraction patterns were analyzed by the Rietveld method using the RIETAN-FP program [24]. The lattice parameters were determined with the silicon 640b used as an internal standard.

3. Results and discussion

In Figure 1, electrical resistivity of $YbGa_xSi_{2-x}$ at low temperatures measured under magnetic fields are shown. Defining T_C^{onset} as the temperature at which the resistivity is 90% of the values at 3 K, T_C^{onset} is estimated to be 2.5 K, 2.32 K, 2.15 K and 1.89 K for $x = 1.15$, 1.18, 1.26 and 1.31, respectively. Under magnetic fields, T_C^{onset} also decreases almost linearly with fields. Field dependence of T_C^{onset} is plotted in Figure 2. The upper critical field at zero temperature, $H_{C2}(0)$, is estimated using the Wertharmer Helfand Hohenberg relation, $H_{C2}(0) = 0.69 \times (-\mathrm{d}H_{C2}/\mathrm{d}T) \times T_C$, which gives $H_{C2}(0) = 5.9$ kOe, 5.3 kOe, 4.2 kOe and 3.1 kOe, for $x = 1.15$, 1.18, 1.26 and 1.31, respectively. The value of $H_{C2}(0)$ for $x= 1.15$ is consistent with that recently obtained from the measurements of electrical resistivity down to 0.3 K under magnetic fields for $YbGa_{1.15}Si_{0.85}$ [25].

[1] This paper is for the proceedings of *'International and Interdisciplinary Workshop on Novel Phenomena in Integrated Complex Sciences: from Non-living to Living Systems'*, held in October 2010, Kyoto, Japan, which was in part supported by the Grant-in-Aid for the Global COE Program, 'The Next Generation of Physics, Spun from Universality and Emergence', from the Ministry of Education, Culture, Sports, Science and Technology (MEXT) of Japan.

Journal of Physics: Conference Series **344** (2012) 012012 doi:10.1088/1742-6596/344/1/012012

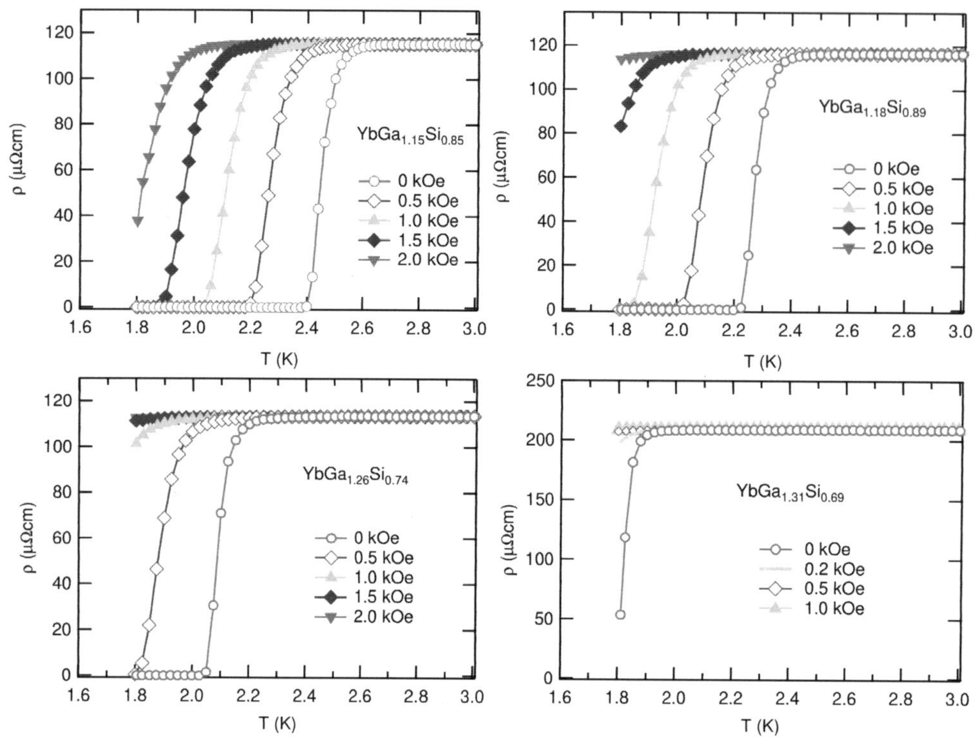

Figure 1. Electrical resistivity measured under magnetic fields of $YbGa_xSi_{2-x}$ with x = 1.15, 1.18, 1.26 and 1.31.

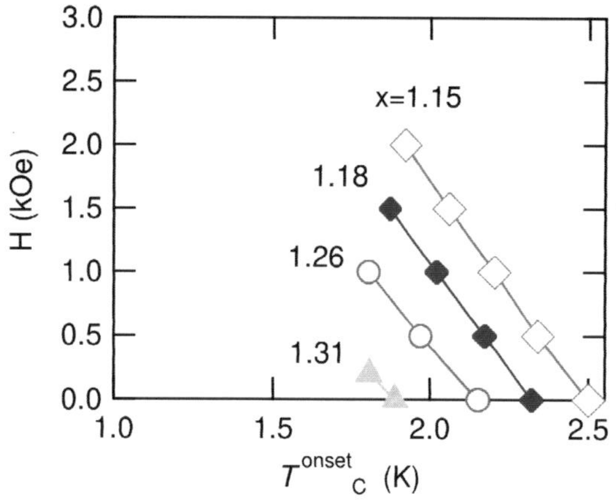

Figure 2. Onset temperature of superconductivity, T_C^{onset}, under magnetic fields for $YbGa_xSi_{2-x}$.

In Figure 3, low temperature specific heat of $YbGa_xSi_{2-x}$ divided by temperature, C/T, is plotted against T^2. For zero magnetic field (Figure 3 (a)), a peak exists at $T \sim 2.3$ K. This peak can be attributed to the superconducting transition of the $YbGa_xSi_{2-x}$ phase and/or the antiferromagnetic transition of Yb_2O_3 impurity, the latter of which is sometimes observed in the low temperature specific heat for Yb-based intermetallic compounds [27, 28].

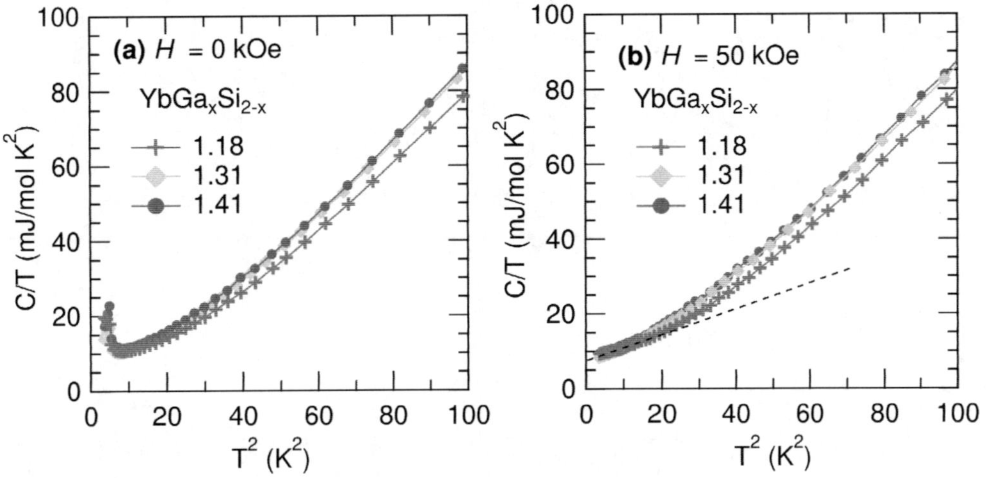

Figure 3. Specific heat divided by temperature, C/T, plotted as a function of squared temperature, T^2 under zero field (a), and under the field of 50 kOe (b). The dashed line in (b) represents the relation $C/T = \gamma + \beta T^2$.

For $x = 1.18$, T_C is close to T_N of Yb_2O_3, 2.3 K, therefore both the effects are considered to be superimposed in the peak of C/T. For $x = 1.31$ and 1.41, T_C is lower than 2.3 K. The peak in C/T for the two compounds at $T = 2.3$ K is then attributed to the antiferromagnetic transition of Yb_2O_3 impurity. The entropy corresponding to the antiferromagnetic transition is calculated by integrating C/T over T down to 0 K, for which C/T values at temperatures below 1.8 K were extrapolated assuming the magnon contribution, $C/T \propto T^2$. The entropy due to Yb_2O_3 thus estimated is $0.0034R\ln 2$, where R is the gas constant. This indicates the amount of Yb_2O_3 is less than 0.4% of total Yb. Hence, the effect of Yb_2O_3 to the superconducting properties in $YbGa_xSi_{2-x}$ is not important. Nevertheless, the effect of antiferromagnetic transition of Yb_2O_3 makes it difficult to estimate the intrinsic electronic specific-heat coefficient, γ, of $YbGa_xSi_{2-x}$ at low temperatures. We then applied magnetic field of 50 kOe to suppress the effect of Yb_2O_3, and measured the specific heat. The results are shown in Figure 3 (b). In this figure, the anomaly due to Yb_2O_3 is not seen, thereby we are able to estimate the electronic specific heat coefficient by fitting using the relation $C/T = \gamma + \beta T^2$. The values of γ are obtained to be $\gamma \simeq 8$ mJ/molK^2 and are almost independent to chemical composition x.

The value of $\gamma = 8$ mJ/molK^2 indicates that the electron-electron correlation in this compounds is not important. This suggests the normal BCS-type superconductivity for the present compound. Nevertheless, it is notable that the value $\gamma = 8$ mJ/molK^2 is slightly larger than those for CaGaSi ($\gamma = 4$ mJ/molK^2) and SrAlSi ($\gamma = 5.4$ mJ/molK^2) [25]. This suggests that the $4f$ electrons participate to the formation of the Fermi surface to some extent. The x-ray absorption and photoelectron spectroscopy indeed showed weak mixed valence state, $Yb^{2.3+}$, for the present compound [21]. In addition, almost constant γ over x means the extent of hybridization between $4f$ and the conduction electrons does not change with chemical composition. Again, this is in good agreement with the results of x-ray emission spectroscopy, which showed the valence of Yb is almost constant, $Yb^{2.3+}$ over x, pointing to the same order of hybridization [19].

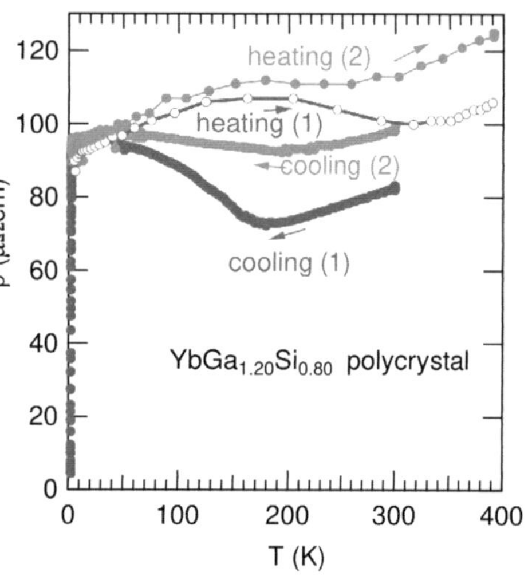

Figure 4. Temperature dependence of electrical resistivity for $YbGa_{0.8}Si_{1.2}$.

In the course of electrical resistivity measurements, we observed an irreversible behavior in the temperature dependence of the resistivity, $\rho(T)$. For some samples, $\rho(T)$ shows a hump-like increase with decreasing temperature, while it shows rather a monotonic increase with increasing temperature. Figure 4 displays $\rho(T)$ of $YbGa_{0.8}Si_{1.2}$, which showed the most prominent irreversible behavior among our measurements.

$\rho(T)$ of this sample shows a minimum at $T = 180$ K and increases with decreasing temperature for the first cooling process. Then the measurement on the heating process shows a broad maximum around $T = 200$ K. The measurement of $\rho(T)$ for the second time again shows similar anomaly, but those are less pronounced compared to the first time. Possible mechanisms for these anomalous temperature dependence of $\rho(T)$ include electronic phase transition, such as valence transition of Yb, charge density wave (CDW) formation, or structural deformation, etc. Here, valence transition of Yb is ruled out since x-ray spectroscopic measurements revealed that the valence of Yb is independent to temperature [21].

To clarify if there is a structural phase transition at low temperature, powder x-ray diffraction has been taken for $YbGa_{1.2}Si_{0.8}$ at low temperatures. The diffraction patterns at 300 K and 12 K are shown in Figure 5, with the results of Rietveld fittings, analyzed by the RIETAN-FP program [24]. No extra peaks other than those of the AlB_2-type structure and the standard Si have been observed. This rules out the possibility of a structural phase transition. In the AlB_2-type structure, atomic positions are not allowed to move from the symmetric positions. In case of slight structural deformation from the AlB_2-type structure, however, one can expect an enhancement of the isotropic atomic displacement parameter, B. We found that the values of B shows monotonic decreases for both Yb and Ga/Si sites with decreasing temperature, without any anomalous behavior at low temperature. Thus the possibility of structural deformations at low temperature is unlikely, though we should mention that the reliability of this B analysis is not sufficient because of the low signal-to-noise ratio due to the in-house powder x-ray diffraction equipment. It is desired to perform neutron diffraction and/or synchrotron x-ray diffraction experiments to discuss the thermal effect in more detail.

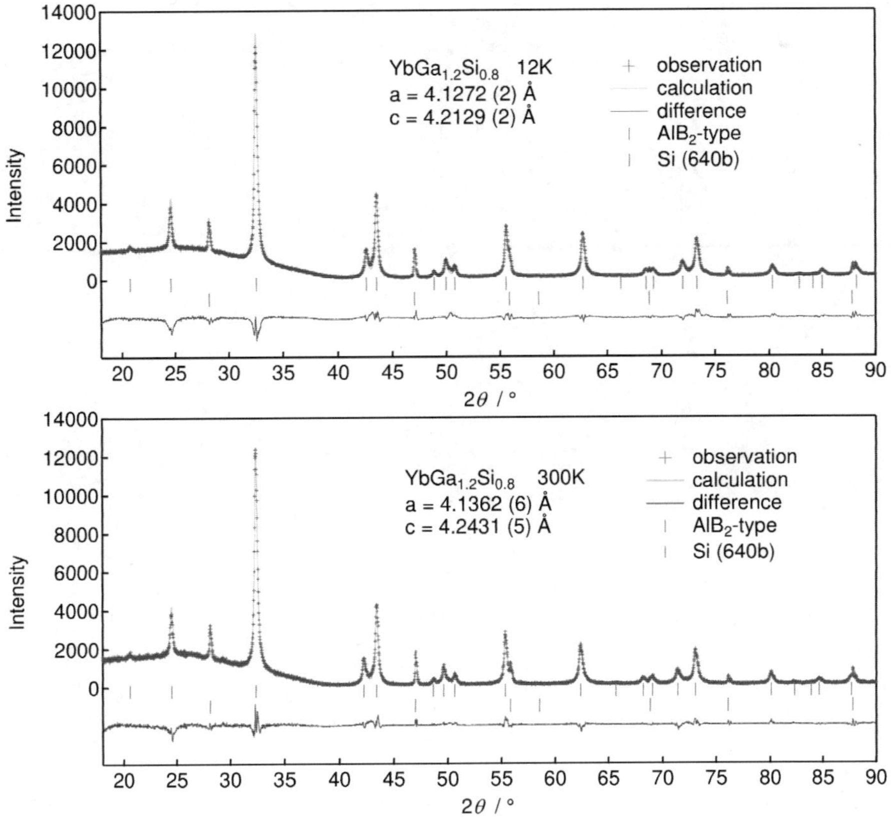

Figure 5. Powder x-ray diffraction patterns and the results of the Rietveld fittings for $YbGa_{1.2}Si_{0.8}$ at T = 12 K and 300 K. Reliable factors are R_{WP} = 9.2% and S = 2.4 for 12 K and R_{WP} = 10.4% and S = 2.6 for 300K.

Figure 6 shows temperature dependence of the lattice parameter, a and c, the unit cell volume V, and the ratio c/a, between 12 and 300 K for $YbGa_{1.2}Si_{0.8}$. These values were determined by the Rietveld fitting of the powder XRD data with Si 640b used as internal standard. Both a and c show monotonic decreases with the decrease of temperature [26]. As a result, the lattice volume V also decreases monotonically.

The lattice parameter a decreases approximately by 0.2% from 300 to 100 K, giving the thermal expansion coefficient $\alpha_a = 1.0\times10^{-5}K^{-1}$. Similarly, c decreases by 0.54% from 300 to 100 K, yielding $\alpha_c = 2.7\times10^{-5}K^{-1}$. These values of α_a and α_c are very close to those of the related compound YbGaGe and $YbGa_xGe_{2-x}$, reported previously [27, 29, 30].

Up to now no structural properties appear to suggest the correlation with the irreversible behavior in the temperature dependence of the electrical resistivity. One possible explanation would be the temperature dependent c/a ratio shown in Fig. 6 (D), which decreases by 0.5% with decreasing temperature from 300 K to 12 K. The monotonic decrease in the c/a ratio can lead increasing lattice strain in the polycrystalline samples, possibly resulting in the large hysteresis behavior in $\rho(T)$. If this is the case, the irreversible $\rho(T)$ would only appear for polycrystalline samples and is not expected to occur for single crystals. Another possible mechanism would be the formation of CDW transition below 100 K. For this case, the temperature at which CDW transition takes place would be sensitive to the carrier density. In our case, the temperature

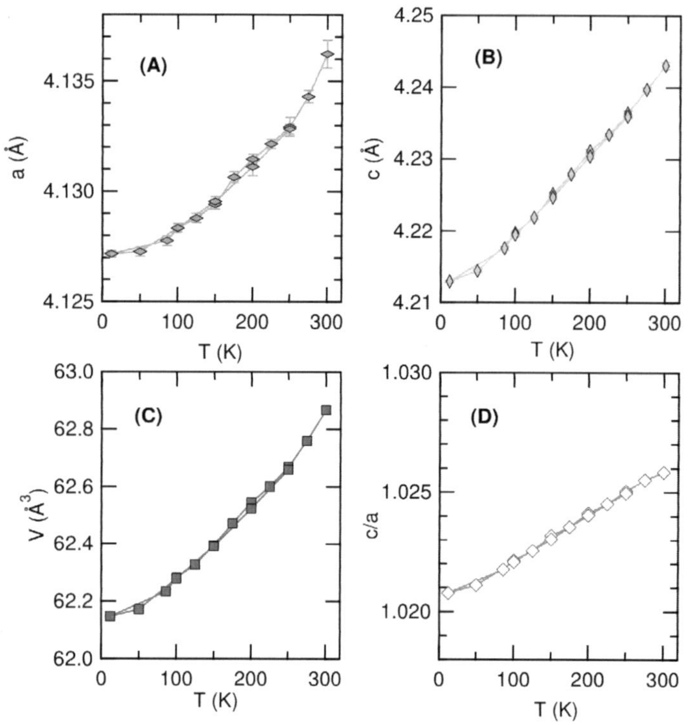

Figure 6. Temperature dependence of the lattice parameters a and c, the unit cell volume V, and the ratio c/a, for $YbGa_{1.2}Si_{0.8}$ annealed sample.

where the irreversible behavior in $\rho(T)$ occurs is found to shift to lower temperatures for larger x samples, which corresponds to the decrease of valence electron number. Therefore, more detailed studies of the electronic structure and the lattice dynamics are desired for this $YbGa_xSi_{2-x}$ system.

4. Summary

Electrical resistivity, specific heat and thermal expansion have been investigated for the new superconductor $YbGa_xSi_{2-x}$. T_C shows a linear decrease with increase of x. The estimated upper critical field at zero temperature decreases accordingly from 5.9 kOe for $x = 1.15$ to 3.1 kOe for $x = 1.31$. The results of specific heat measurements show that the electronic specific heat coefficients are almost independent to chemical composition x, with $\gamma \simeq 8$ mJ/molK^2 for all x. This small value of γ indicates that electronic correlation in $YbGa_xSi_{2-x}$ is not enhanced, and is consistent with the nearly divalent Yb valence, $Yb^{2.3+}$, observed by the x-ray spectroscopic measurements. Temperature dependence of the electrical resistivity reveals an irreversible behavior. This $\rho(T)$ behavior suggests some structural deformation occurs at low temperature. On the other hand, no structural phase transition is observed by powder x-ray diffraction for $YbGa_{1.2}Si_{0.8}$. Temperature dependence of the lattice parameters shows an monotonic decrease with decreasing temperature, both for a and c. Further detailed studies are needed to address the electronic properties of the $YbGa_xSi_{2-x}$ system, including the possibility of the CDW formation at low temperature.

Acknowledgment

We thank Takao Furubayashi for allowing us to use the low-temperature x-ray diffraction equipments. We thank Yuko Isozaki and Toshiya Ogiwara of Materials Analysis Station, NIMS, for the EPMA measurements. One of the authors (NT) thanks Hitoshi Yamaoka and Ignace Jarrige for a lot of helpful discussion. He also acknowledges the Yukawa Institute for Theoretical Physics at Kyoto University, for the financial support to his participation to 'International and Interdisciplinary Workshop on Novel Phenomena in Integrated Complex Sciences: from Non-Living to Living Systems'.

References

[1] Imai M, Abe E, Ye J, Nishida K, Kimura T, Honma K, Abe H and Kitazawa H 2001 *Phys. Rev. Lett.* **87** 077003
[2] Imai M, Nishida K, Kimura T and Abe H 2002 *Appl. Phys. Lett.* **80** 1019
[3] Imai M, Nishida K, Kimura T and Abe H 2002 *Physica C* **377** 96
[4] Imai M, Nishida K, Kimura T, Kitazawa H, Abe H, Kito H and Yoshii K 2002 *Physica C* **382** 361
[5] Lorenz B, Lenzi J, Cmaidalka J, Meng R L, Sun Y Y, Xue Y Y and Chu C W 2002 *Physica C* **383** 191
[6] Meng R L, Lorenz B, Wang Y S, Cmaidalka J, Sun Y Y, Xue Y Y, Meen J K and Chu C W 2002 *Physica C* **382** 113
[7] Meng R L, Lorenz B, Cmaidalka J, Wang Y S, Sun Y Y, Lenzi J, Meen J K, Xue Y Y and Chu C W 2003 *IEEE Transact. Appl. Supercond.* **13** 3042
[8] Yamanaka S, Otsuki T, Ide T, Fukuoka H, Kumashiro R, Rachi T, Tanigaki K, Guo F Z and Kobayashi K 2007 *Physica C* **451** 19
[9] Nagamatsu J, Nakagawa N, Muranaka T, Zenitani Y and Akimitsu J 2001 *Nature* **410** 63
[10] Lorenz B, Meng R L, Xue Y Y and Chu C W 2001 *Phys. Rev. B* **64** 052513
[11] Kang W N, Jung C U, Kim K H P, Park M S, Lee S Y, Kim H H, Choi E M, Kim K H, Kim M S and Lee S I 2001 *Appl. Phys. Lett.* **79** 982
[12] Choi H J, Roundy D, Sun H, Cohen M L and Louie S G 2002 *Nature* **418** 758
[13] Shukla A, Calandra M, d'Astuto M, Lazzeri M, Mauri F, Bellin C, Krisch M, Karpinski J, Kazakov S M, Jun J, Daghero D and Parlinski K 2003 *Phys. Rev. Lett.* **90** 095506
[14] Baron A Q R, Uchiyama H, Tanaka Y, Tsutsui S, Ishikawa D, Lee S, Heid R, Bohnen K P, Tajima S and Ishikawa T 2004 *Phys. Rev. Lett.* **92** 197004
[15] Imai M, Sadki E H S, Abe H, Nishida K, Kimura T, Sato T, Hirata K and Kitazawa H 2003 *Phys. Rev. B* **68** 064512
[16] Kuroiwa S, Baron A Q R, Muranaka T, Heid R, Bohnen K P and Akimitsu J 2008 *Phys. Rev. B* **77** 140503
[17] Heid R, Bohnen K P, Renker B, Adelmann P, Wolf T, Ernst D and Schober H 2007 *J. Low Temp. Phys.* **147** 375
[18] Imai M, Sato A, Aoyagi T, Kimura T, Matsushita Y and Tsujii N 2008 *J. Am. Chem. Soc.* **130** 2886
[19] Tsujii N, Imai M, Yamaoka H, Jarrige I, Oohashi H, Tochio T, Handa K, Ide J, Atsuta H, Ito Y, Yoshikawa H and Kitazawa H 2010 *Chem. Mater.* **22** 4690
[20] Ohmura A, Fujimaki K, Ishikawa F, Yamada Y, Tsujii N and Imai M 2011 *Phys. Rev. B* **84** 104520
[21] Yamaoka H, Jarrige I, Tsujii N, Imai M, Lin J F, Matsunami M, Eguchi R, Arita M, Shimada K, Namatame H, Taniguchi M, Taguchi M, Senba Y, Ohashi H, Hiraoka N, Ishii H and Tsuei K D 2011 *Phys. Rev. B* **83** 104525
[22] Ebihara T, Inada Y, Murakawa M, Uji S, Terakura C, Terashima T, Yamamoto E, Haga Y, Ōnuki Y and Harima H 2000 *J. Phys. Soc. Japan* **69** 895
[23] Ebihara T, Uji S, Terakura C, Terashima T, Yamamoto E, Haga Y, Inada Y and Ōnuki Y 2000 *Physica B* **281-282** 754
[24] Izumi F and Momma K 2007 *Solid State Phenom.* **130** 15
[25] Imai M, Tsujii N, Tachibana M and Uji S 2011 *Superconduct. Sci. Technol.* **24** 055015
[26] An anomaly in the temperature dependence of a below 100 K was suggested at the Workshop, but it turned out to be errorneous.
[27] Drymiotis F R, Lee Y, Lawes G, Lashley J C, Kimura T, Shapiro S M, Migliori A, Correa V and Fisher R A 2005 *Phys. Rev. B* **71** 174304
[28] Görlach T, Pfleiderer C, Grube K and Löhneysen H v 2005 *Phys. Rev. B* **71** 033101
[29] Muro Y, Nakagawa T, Umeo K, Itoh M, Suzuki T and Takabatake T 2004 *J. Phys. Soc. Japan* **73** 1450
[30] Tsujii N, Furubayashi T, Kitazawa H and Kido G 2005 *J. Alloys Compd.* **393** 41

Journal of Physics: Conference Series **344** (2012) 012013 doi:10.1088/1742-6596/344/1/012013

Optical conductivity of $Ca_{1-x}Na_xV_2O_4$ ($x = 0, 2/3$ and 1)

Hiroya Sakurai[1,*], Akinori Irizawa[2,3], and Takao Nanba[2]

[1]National Institute for Materials Science, 1-1 Namiki, Tsukuba, Ibaraki 305-0044, Japan
[2]Graduate School of Science and Technology, Kobe University, 1-1 Rokkodai, Nada, Kobe, Hyogo, 657-8501, Japan

E-mail: `sakurai.hiroya@nims.go.jp`

Abstract. Optical conductivity was measured for CaV_2O_4, $Ca_{1/3}Na_{2/3}V_2O_4$ and NaV_2O_4. The spectra of CaV_2O_4 showed insulating behavior with the band gap of 0.07 eV, which agrees well with the energy gap estimated from electrical resistivity data. On the other hand, in the spectra of NaV_2O_4, Drude peak was clearly observed at 300 K and 8 K, supporting the metallic nature below and above antiferromagnetic transition at 140 K. For $Ca_{1/3}Na_{2/3}V_2O_4$, the spectrum at 300 K showed Drude peak, while that at 8 K showed insulating behavior with the band gap of 0.08 eV. This change is caused by metal-insulator transition observed by the resistivity measurement.

1. Introduction

Both CaV_2O_4 and NaV_2O_4 crystallize in calcium ferrite type structure, which is shown in Fig. 1, but their physical properties are contrastive to each other. CaV_2O_4 is electrically insulating, and shows antiferromagnetic transition at $T_N = 78$ K [1]; the temperature apparently depends on the samples probably due to calcium deficiency [2]. Magnetic susceptibility shows a broad maximum at around 300 K because of one dimensional spin correlation and/or spin frustration in the double chains of V_2O_4. Magnetic structure was determined by detailed NMR and neutron diffraction measurements [1, 3]; magnetic moments are aligned almost perpendicular to the chain directions. On the other hand, NaV_2O_4 shows antiferromagnetic transition at $T_N = 140$ K maintaining metallic conductivity [4]. The antiferromagnetic phase is composed, at least, of three kinds of subphases, whose boundary is very sensitive to temperature and magnetic field [5]. According to μSR, Neutron diffraction, and NMR measurement, magnetic moments in the ground state are helically rotating with incommensurate propagation vector (0 0.191 0) [6, 7, 8]. The purpose of our study is to make clear the physical properties which appear in the solid solution system $Ca_{1-x}Na_xV_2O_4$.

This workshop was supported in part by the Grant-in-Aid for the Global COE Program "The Next Generation of Physics, Spun from Universality and Emergence" from the Ministry of Education, Culture, Sports, Science and Technology (MEXT) of Japan.

Journal of Physics: Conference Series **344** (2012) 012013 doi:10.1088/1742-6596/344/1/012013

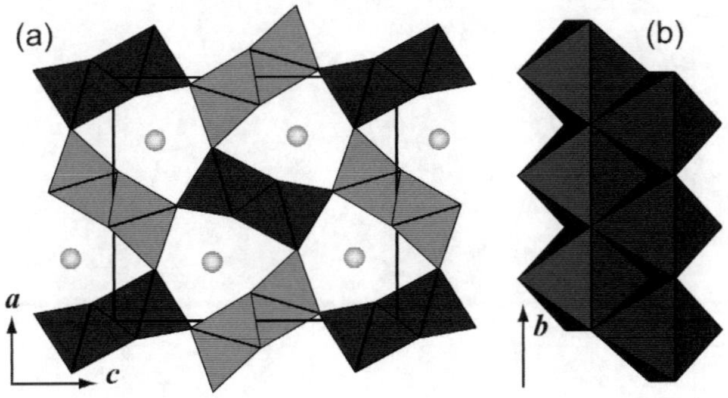

Figure 1. Structure of AV_2O_4 (A = Ca or Na) (a) and the double chain (b) The octahedra and the circle represent VO_6 octahedra and A, respectively.

2. Experimental

CaV_2O_4 was synthesized in flowing H_2 gas at 1000°C for 6 hours from CaV_2O_6, which was prepared from a stoichiometric mixture of $CaCO_3$ and V_2O_5 at 700°C in air. The powder sample obtained was pressed under high pressure of 6 GPa using a belt-type press and heated at 1300°C for 1 hour to obtain a hard pellet for optical conductivity measurement. On the other hand, $Ca_{1/3}Na_{2/3}V_2O_4$ and NaV_2O_4 were made by high pressure synthesis technique from stoichiometric mixtures of CaV_2O_4, $Na_4V_2O_7$ and V_2O_3. The mixtures were sealed in Au capsules in a dry box filled with Ar gas because $Na_4V_2O_7$ is reactive with moisture in air. Then,

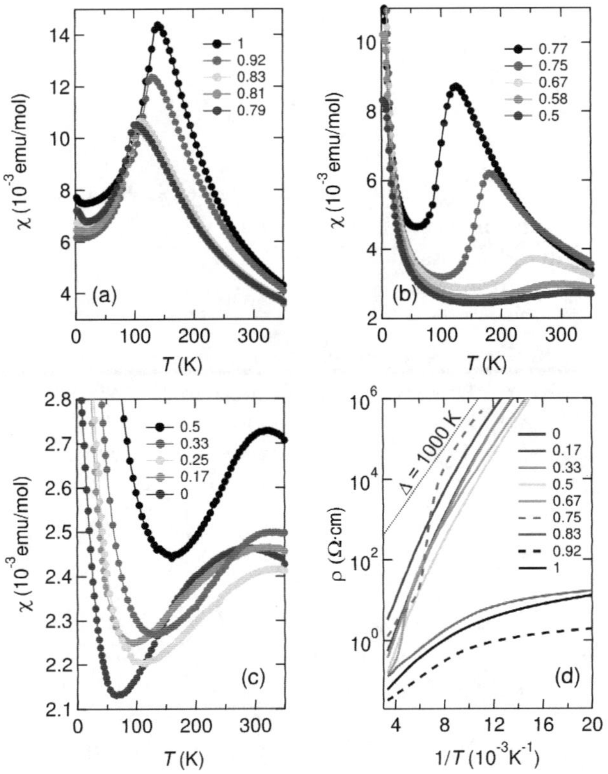

Figure 2. Temperature dependence of magnetic susceptibility (*a*, *b*, *c*) and electrical resistivity (*d*) of $Ca_{1-x}Na_xV_2O_4$. The dotted line in panel d shows the slope of $\Delta = 1000$ K.

[3] Present address: The Institute of Science and Industrial Research, Osaka University, 8-1 Mihogaoka, Ibaraki, Osaka, 567-0047, Japan

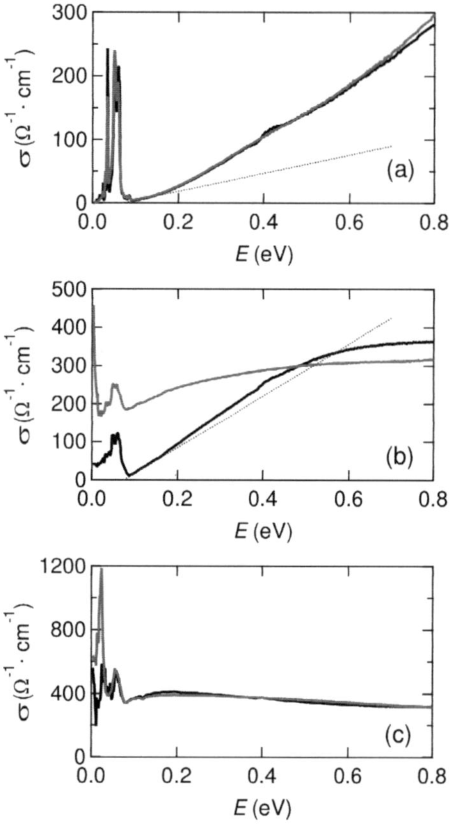

Figure 3. Optical conductivity of CaV_2O_4 (*a*), $Ca_{1/3}Na_{2/3}V_2O_4$ (*b*), and NaV_2O_4 (*c*) as a function of photon energy. In each panel, the black and red lines represent the data at 8 K and 300 K, respectively. The blue dotted lines in panel *a* and *b* are guides to eye to show the position of the edges of the interband excitation.

the samples were pressed under 6 GPa and heated at 1300°C for 1 hour. After the synthesis, they were quenched to room temperature and then the pressure was gradually released to ambient pressure. $Na_4V_2O_7$ was prepared from a stoichiometric mixture of $NaCO_3$ and V_2O_5 by heating it at 550°C in air with intermediate grindings. V_2O_3 was obtained by reduction of V_2O_5 by flowing H_2 gas.

Magnetic susceptibility data were collected under magnetic field of 1 T using a commercial magnetometer, MPMS-XL (Quantum Design). Electrical resistivity was measured by conventional four probe method in PPMS (Quantum Design). Optical conductivity was obtained from reflectivity of wide energy range by Kramers-Kronig analysis. The reflectivity was measured at 8 K and 300 K using an FT-IR spectrometer with conventional light sources for the photon energy up to 1.5 eV.

3. Results and Discussion

Figure 2 shows magnetic susceptibility and electrical resistivity of $Ca_{1-x}Na_xV_2O_4$. The peak of the susceptibility of NaV_2O_4 corresponds to the antiferromagnetic transition [6, 7, 9], and the peak temperature decreases with decreasing x down to $x = 0.78$. At this x range, electrical resistivity shows no significant change as a function of chemical composition or temperature. Thus, the metallic antiferromagnetic state is preserved for $x > 0.78$, which is supported by other experiments [6, 7, 8, 9]. On the other hand, as the x value further decreases down to about 0.3, the peak temperature of magnetic susceptibility, $T_{\rm max}$, rapidly increases and reaches around 300 K. In addition, electrical resistivity markedly increases below $T_{\rm max}$, suggesting metal-insulator transition. Thus, electronic ground state for $\sim 0.3 < x < 0.78$ is completely different from that for $x > 0.78$. Below $x \sim 0.3$, $T_{\rm max}$ slightly decreases as seen in Fig. 2*c*, but no clear metal-insulator transition was observed below 400 K any longer. The band gap for $x < 0.78$ was roughly estimated to be $\Delta \sim 1000$ K from the slope of Arrhenius plot.

Because electronic state of $Ca_{1-x}Na_xV_2O_4$ is separated into three kinds depending on the x value, optical conductivity was measured for the $x = 0$, 2/3 and 1 samples at 8 K and 300 K. The spectra obtained are shown in Fig. 3. In all of them, the sharp peaks were observed around $E = 0.05$ eV, which originate from phonons. For NaV_2O_4 ($x = 1$), finite spectral weight is seen at $E = 0$ eV at both 8 K and 300 K, which is due to Drude contribution and is consistent with the metallic behavior below and above T_N. On the other hand, the spectra at 8 K and 300 K for CaV_2O_4 ($x = 0$) have no Drude term, indicating insulating nature of it. The conductivity starts increasing at $E = 0.07$ eV, which agrees well with Δ (0.086 eV). Thus, the increase is due to interband excitation. For $Ca_{1/3}Na_{2/3}V_2O_4$ ($x = 2/3$), the spectrum at 300 K is similar to that of NaV_2O_4, showing metallic nature, while that at 8 K clearly indicates that the band gap opens as in the case of CaV_2O_4. The spectral change is due to metal-insulator transition seen in the resistivity. The band gap was estimated to be 0.08 eV, from the edge of the broad peak around $E = 0.7$ eV.

4. Conclusion

Magnetic susceptibility and electrical resistivity were measured for $Ca_{1-x}Na_xV_2O_4$, and it was proved that electronic state is separated into three kinds depending the x value. For $x > 0.78$, the metallic antiferromagnetic ordered state occurs, for $x <\sim 0.3$ the insulating behavior was observed at the entire temperature region below 400 K, and for $\sim 0.3 < x < 0.78$, the metal-insulator transition appears. These were confirmed by the optical conductivity measurement. The band gap of the insulating state for $x = 0$ and 2/3 was estimated to be 0.07 eV and 0.08 eV, respectively, which agree well with the slopes of Arrhenius plot of the resistivity.

Acknowledgments

Special thanks to Dr. Taniguchi (NIMS) for his support for high pressure synthesis. This work was supported in part by Grant-in-Aid for Scientific Research (A) from Japan Society for the Promotion of Science (JSPS) (22246083), by Grant-in-Aid for Scientific Research from MEXT (19560665), and by FIRST from JSPS. The authors thank the Yukawa Institute for Theoretical Physics at Kyoto University, where this work was completed during the YITP-W-10-12 on "International and Interdisciplinary Workshop on Novel Phenomena in Integrated Complex Sciences: from Non-living to Living Systems".

References

[1] Zong X, Suh B J, Niazi A, Yang J Q, Schlagel D L, Lograsso T A and Johnston D C, 2008 *Phys. Rev.* B **77** 014412
[2] Sakurai H, Takeya H and Hirata K, *J. Phys.: Conf. Ser.* in press
[3] Pieper O, Lake B, Daoud-Aladine A, Reehuis M, Prokěs K, Klemke B, Kiefer K, Yan J Q, Niazi A, Johnston D C and Honecker A, 2009 *Phys Rev* B **79** 180409
[4] Yamaura K, Arai M, Sato A, Karki A B, Young D P, Movshovich R, Okamoto S, Mandrus D and Takayama-Muromachi E, 2007 *Phys. Rev. Lett.* **99** 196601
[5] Sakurai H 2008 *Phys. Rev.* B **78** 094410
[6] Sugiyama J, Ikedo Y, Goko T, Ansaldo E J, Brewer J H, Russo P L, Chow K H, and Sakurai H 2008 *Phys. Rev* B **78** 224406
[7] Nozaki H, Sugiyama J, Månsson M, Harada M, Pomjakushin V, Sikolenko V, Cervellino A, Roessli B and Sakurai H 2010 *Phys. Rev.* B **81** 100410
[8] Takeda H and Itoh M, private communication
[9] Takeda H, Itoh M and Sakurai H 2010 *J. Phys.: Conf. Ser.* **200** 012200

Journal of Physics: Conference Series **344** (2012) 012015 doi:10.1088/1742-6596/344/1/012015

Effect of dimensionality crossover on magnetovolume properties of quasi one-dimensional weakly antiferromagnetic metals

N. Hatayama[1], R Konno[1], Y Takahashi[2]
[1]Kinki University Technical College, 2800 Arima-cho, Kumano, Mie 519-4395, Japan
[2]Graduate School of Material Science, University of Hyogo, Kouto, Kamigori, Ako 678-1297, Japan

E-mail: hatayama@ktc.ac.jp and r-konno@ktc.ac.jp

Abstract. Magnetic properties and thermal expansion of quasi one-dimensional itinerant electron magnets close to the antiferromagnetic instability are studied. By controlling the relative spectral dispersions of spin fluctuation spectra depending on the direction of crystal axes, we show that the three dimensional properties are changed with increasing the one-dimensional anisotropy. Crossover behaviors are discussed on the temperature dependence of the magnetic susceptibility and the magnetovolume expansion around their critical regions.

1. Introduction

Since the spin fluctuation theory of magnetovolume effect was published in 1980 by Moriya and Usami [1], the volume change of crystals from the magnetic origin have been usually analyzed by the following formula:

$$\omega \equiv \frac{\delta V}{V} = \rho K C[M^2(T) + \xi^2(T)], \quad \xi^2(T) = \sum_q < \delta \mathbf{M}_q \cdot \delta \mathbf{M}_{-q} >, \tag{1}$$

where $M(T)$, $\delta\mathbf{M}_q$, and $\xi(T)$ represent the magnetization, the wave-vector dependent magnetic fluctuation and the thermal spin fluctuation amplitude. The magnetovolume coupling constant, the compressibility, and a density of magnetic atoms are, respectively, denoted by C, K, and $\rho = N_0/V$, where N_0 and V are the number of magnetic atoms and the volume of the crystal. The result was derived by extending the Stoner-Wohlfarth-Edwards free energy [2] of the magnetovolume effect, simply by including the effect of thermal spin fluctuations. Though (1) looks reasonable, it violates the thermodynamic consistency because Grüneisen's relation between the magnetic specific heat and the thermal expansion coefficient is not satisfied at low temperatures, as was pointed out by Takahashi and Nakano [3].

Recently, the inconsistency was resolved by Takahashi and Nakano [3, 4]. Their thermal volume expansion is derived from the explicit volume dependence of the same free energy that is used in their treatment of the magnetic specific heat. Grüneisen's relation is, therefore, automatically satisfied at low temperatures. According to them, the thermal expansion in itinerant electron ferromagnets is dominated by the T^2-linear behavior at low temperatures, and the T-linear slope of the thermal expansion coefficient rapidly increases towards the magnetic instability point. Similar enhancement of the thermal expansion coefficients has been also predicted [5] for itinerant nearly and weakly antiferromagnetic metals near the instability point.

A recent study on thermal expansion coefficients of lower-dimensional itinerant weak antiferromagnets shows even stronger enhancement around the magnetic instability point [5]. It is, however, very difficult to find strictly low-dimensional materials in nature. Only usually found are quasi two-dimensional or one-dimensional systems. The purpose of this paper is, therefore, to study the magnetic properties and the thermal expansion of quasi one-dimension exchange enhanced paramagnets (P) and antiferromagnets (AF) close to the instability points. Our particular concerns are the crossover from three-dimensional (3-D) to one-dimensional (1-D) behaviors on the temperature dependence of the above magnetic and magnetovolume properties.

In what follows in section 2, we show how magnetic properties of quasi one-dimensional itinerant magnets close to the antiferromagnetic instability are predominantly determined by the effect of spin fluctuations. Based on the results, the temperature dependence of the magneto-thermal expansion is discussed in section 3, followed by concise summaries in section 4.

2. Magnetic properties of quasi 1-D itinerant magnets near the AF instability

Let us first discuss magnetic properties of itinerant quasi 1-D antiferromagnets based on the spin amplitude conservation by Takahashi [6]. It means that the local spin amplitude defined by

$$\langle \mathbf{S}_{loc}^2 \rangle = \frac{3}{N_0^2} \sum_q \int_0^\infty \frac{\mathrm{d}\nu}{\pi} \coth(\nu/2k_BT) \mathrm{Im}\chi(\mathbf{Q}+\mathbf{q},\nu), \tag{2}$$

is almost conserved independent of the temperature and the presence of magnetic field. The total amplitude in (2) is divided into thermal and zero-point amplitudes defined by

$$\begin{aligned} 3A_t(\kappa_s^2, T) &\equiv \langle \mathbf{S}_{loc}^2 \rangle_\mathrm{t} = \frac{6}{N_0^2} \sum_q \int_0^\infty \frac{\mathrm{d}\nu}{\pi} \frac{1}{\mathrm{e}^{\nu/k_BT}-1} \mathrm{Im}\chi(\mathbf{Q}+\mathbf{q},\nu), \\ 3A_z(\kappa_s^2) &\equiv \langle \mathbf{S}_{loc}^2 \rangle_\mathrm{z} = \frac{3}{N_0^2} \sum_q \int_0^\infty \frac{\mathrm{d}\nu}{\pi} \mathrm{Im}\chi(\mathbf{Q}+\mathbf{q},\nu). \end{aligned} \tag{3}$$

The imaginary part of the dynamical susceptibility for quasi 1-D itinerant magnets near the antiferromagnetic instability point is assumed to be given by the double Lorentzian form around the antiferromagnetic wave vector $\mathbf{Q}$:

$$\mathrm{Im}\chi(\mathbf{Q}+\mathbf{q},\nu) = \chi(\mathbf{Q}+\mathbf{q},0) \frac{\nu\Gamma_{\mathbf{Q}+\mathbf{q}}}{\nu^2+\Gamma_{\mathbf{Q}+\mathbf{q}}^2}, \tag{4}$$

where Q is of the order of the zone-boundary vector q_B and the wave-vector dependence of the static magnetic susceptibility and the damping are given by

$$\chi(\mathbf{Q}+\mathbf{q},0) = \frac{\chi(\mathbf{Q})\kappa_s^2}{\kappa_s^2+q_z^2+\varepsilon_a q^2}, \quad \Gamma_{\mathbf{Q}+\mathbf{q}} = \Gamma_{\mathbf{Q}}(\kappa_s^2+q_z^2+\varepsilon_a q^2). \tag{5}$$

In (5), $\kappa_s^2 \propto \chi^{-1}(\mathbf{Q},0)$ stands for the correlation wave-vector squared around the $\mathbf{Q}$ vector, $q^2 = q_x^2 + q_y^2$, and a parameter ε_a introduces an anisotropic wave vector dependence in the system, i.e., a quasi-one dimensionality. As measures of dispersions of the spectrum in $\mathbf{q}$, ν spaces, two spectral widths are defined by

$$T_0 = \frac{\Gamma_0 Q^2}{2\pi}, \quad \frac{N_0}{\chi(\mathbf{0},0)} = \frac{N_0}{\chi(\mathbf{Q},0)}\left[1+\frac{Q^2}{\kappa_s^2}\right] \simeq \frac{N_0 Q^2}{\chi(\mathbf{Q},0)\kappa_s^2} = 2T_A, \tag{6}$$

from differences of $\chi(\mathbf{q},0)^{-1}$ and $\Gamma_{\mathbf{q}}$ for $\mathbf{q} = (0,0,Q)$ and $(0,0,0)$ along q_z-axis. In the same way, the dispersions in the q_x, q_y plane are characterized by

$$\frac{\Gamma_0 \varepsilon_A Q^2}{2\pi} = \varepsilon_A T_0, \quad \frac{N_0}{\chi(\mathbf{0},0)} \simeq \frac{N_0 \varepsilon_A Q^2}{\chi(\mathbf{Q},0)\kappa_s^2} = 2\varepsilon_A T_A. \tag{7}$$

from the differences between those for $\mathbf{q} = (\sqrt{\varepsilon_A}Q, 0, 0)$ and $(0,0,0)$.

2.1. Temperature dependence of magnetic susceptibility

The total spin amplitude conservation can be explicitly written as

$$A_t(\kappa_s^2, T) + A_z(\kappa_s^2) = \begin{cases} A_t(0, T_N) + A_z(0), & \text{for AF} \\ A_z(0) + [A_z(\kappa_{s0}^2) - A_z(0)], & \text{for P} \end{cases} \tag{8}$$

We define κ_{s0} as κ_s in the ground state. The right hand side of (8) represents the amplitude for antiferromagnets at $T = T_N$ or for paramagnets in the ground state, respectively. In the following, the above amplitudes, $A_t(y,t)$ and $A_z(y)$, are regarded as functions of reduced parameters $y = \kappa_s^2/Q^2$ and $t = T/T_0$ as defined below.

$$\begin{aligned} A_t(y,t) &= \frac{T_0 t}{T_A \varepsilon_a} \int_0^1 \mathrm{d}z [\phi(B(z,y)/t) - \phi(B(z, y+\varepsilon_a)/t)] \\ \phi(x) &= -(x - \frac{1}{2}) \ln x + x + \ln \Gamma(x) - \ln \sqrt{2\pi}, \quad B(z,y) = z^2 + y \\ A_z(y) &\simeq A_z(0) - \frac{T_0}{T_A} c(\varepsilon_a) y, \quad c(\varepsilon) = \frac{1}{\varepsilon} \log(1+\varepsilon) + \frac{2}{\sqrt{\varepsilon}} \tan^{-1}(\frac{1}{\sqrt{\varepsilon}}) \end{aligned}$$

After substitution of the y-dependence of $A_z(y)$ for $y \ll 1$, both equations in (8) can be written in the same form:

$$y = y_0 + d(\varepsilon_a) t \int_0^1 \mathrm{d}z \, [\phi(B(z,y)/t) - \phi(B(z, y+\varepsilon_A)/t)], \quad d(\varepsilon) = \frac{1}{\varepsilon c(\varepsilon)}. \tag{9}$$

The first constant y_0 in the right hand side is the ground state value of y for P and is positive. It is negative for AF and is given by the thermal amplitude at $T = T_N$:

$$y_0 = -d(\varepsilon_a) t_N \int_0^1 \mathrm{d}z \, [\phi(B(z,0)/t_N) - \phi(B(z,\varepsilon_a)/t_N)], \tag{10}$$

where $t_N = T_N/T_0$.

From the amplitude conservation (8) in the ground state, we can find the magnetic isotherm there under the presence of the staggered external magnetic field, giving $\sigma_s^2 = 20 A_t(0, T_N)$. It follows that the relation,

$$\sigma_s^2 = 20 A_t(0, T_N) = \frac{20 T_0 t_N}{T_A \varepsilon_a} \int_0^1 \mathrm{d}z \, [\phi(B(z,0)/t_N) - \phi(B(z,\varepsilon_a)/t_N)], \tag{11}$$

between the spontaneous moment σ_s and t_N is satisfied. In the 1-D limit, it is given by

$$\sigma_s^2 = 20 A_t(0, T_N) \simeq \frac{20 T_0}{T_A} \frac{\pi t_N}{\sqrt{\varepsilon_a}}, \quad (\varepsilon_a \to 0). \tag{12}$$

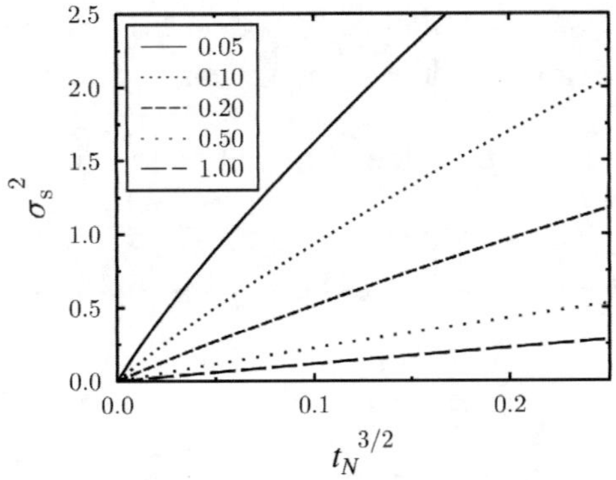

Figure 1. The staggered spontaneous moment squared σ_s^2 versus t_N for $\varepsilon_a = 0.05$, 0.1, 0.2, 0.5, and 1.0 from the top.

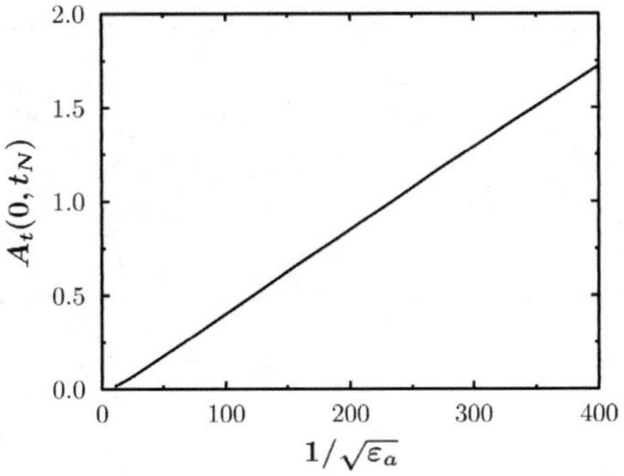

Figure 2. The enhancement of the thermal amplitude at t_N in the 1-D limit.

The thermal amplitude is enhanced proportional to $\varepsilon_a^{-1/2}$ because of the development of the critical fluctuations by lowering the spacial dimensionality of the system. We show in Figure 1, the t_N-dependence of the thermal amplitude in the right hand side of (11) at $t = t_N$. The $t^{3/2}$-linear behaviors for 3-D antiferromagnets observed for larger ε_a are found to crossover to the t_N-linear dependence in the vanishing ε_a limit. In Figure 2, the thermal amplitude is plotted against $\varepsilon_a^{-1/2}$ in agreement with (12).

For paramagnets near the instability, the temperature dependence of the inverse of the reduced staggered susceptibility y is numerically evaluated by solving (9). In the low temperature limit, the following T^2 dependence is obtained analytically [7]:

$$\begin{aligned} y &\simeq y_0 + \frac{1}{24} d(\varepsilon_a) t^2 \left[\frac{1}{\sqrt{y}} \tan^{-1} \frac{1}{\sqrt{y}} - \frac{1}{\sqrt{y+\varepsilon_a}} \tan^{-1} \frac{1}{\sqrt{y+\varepsilon_a}} \right], \\ &\simeq y_0 + \frac{\pi}{48} d(\varepsilon_a) t^2 \left[\frac{1}{\sqrt{y_0}} - \frac{1}{\sqrt{y_0+\varepsilon_a}} \right] \simeq y_0 + \frac{1}{48\sqrt{\varepsilon_a}} t^2 \begin{cases} y_0^{-1/2}, & (y_0 \ll \varepsilon_a) \\ \varepsilon_a y_0^{-3/2}/2, & (y_0 \gg \varepsilon_a) \end{cases}, \end{aligned} \tag{13}$$

for y, $\varepsilon_a \ll 1$ by using $d(\varepsilon) \simeq \varepsilon^{-1/2}/\pi$ and the asymptotic behavior of $\phi(x)$,

$$\phi(x) \simeq \frac{1}{12x}, \quad \text{for } x \gg 1. \tag{14}$$

The T^2-linear coefficient shows crossover behavior from $(\varepsilon_a y_0)^{-1/2}$ for $y_0 \ll \varepsilon_a$ to $y_0^{-3/2}$ for $\varepsilon_a \ll y_0$ depending on the relative magnitude of y_0 and ε_a.

For antiferromagnets, the thermal amplitude near the critical point ($T_N \lesssim T$) can be written in the form,

$$\langle \mathbf{S}^2 \rangle_t (y, t) \simeq \langle \mathbf{S}^2 \rangle_t (0, t) - \frac{3\pi T_N}{2 T_A \varepsilon_a} \sqrt{y}, \quad \text{for } y < \varepsilon_a. \tag{15}$$

Since the zero-point amplitude with y-linear dependence can be neglected around T_N, the temperature dependence of y can be obtained by

$$\langle \mathbf{S}^2 \rangle_t (0, t) - \frac{3\pi T_N}{2 T_A \varepsilon_a} \sqrt{y} = \langle \mathbf{S}^2 \rangle_t (0, t_N), \quad y = \left(\frac{2 T_A \varepsilon_a}{3\pi T_N} \right)^2 [\langle \mathbf{S}^2 \rangle_t (0, t) - \langle \mathbf{S}^2 \rangle_t (0, t_N)]^2. \tag{16}$$

Journal of Physics: Conference Series **344** (2012) 012015 doi:10.1088/1742-6596/344/1/012015

It means that y is proportional to $(T-T_N)^2$, and its coefficient has a ε_a-linear dependence because of the $\varepsilon_a^{-1/2}$ dependence of the thermal amplitude.

We show in Figure 3 the temperature dependence of y for P at low temperatures for various values of ε_a from 0.05 to 1. The slope decreases with increasing ε_a, for $y_0 < \varepsilon_a$ is satisfied. In Figure 4, the temperature dependence of y for AF is shown as a function of $(t-t_N)^2$. In contrast to Figure 3, the slope increases with increasing ε_a because of the ε_a-linear slope for AF.

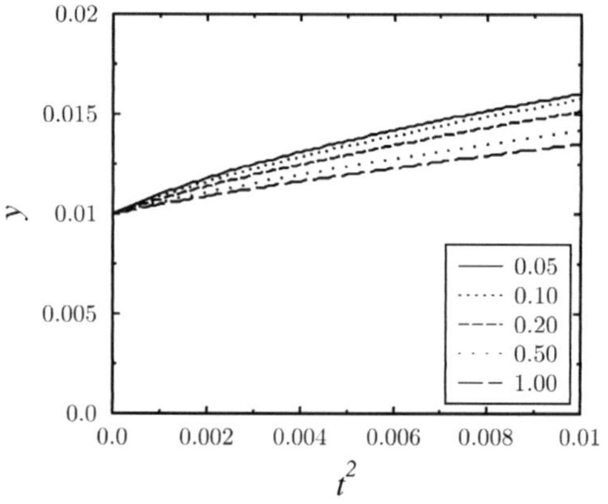

Figure 3. Temperature dependence of y for P with $y_0 = 0.01$ and $T_0/T_A = 0.1$ as a function of t^2. Values of ε_a are shown in the key.

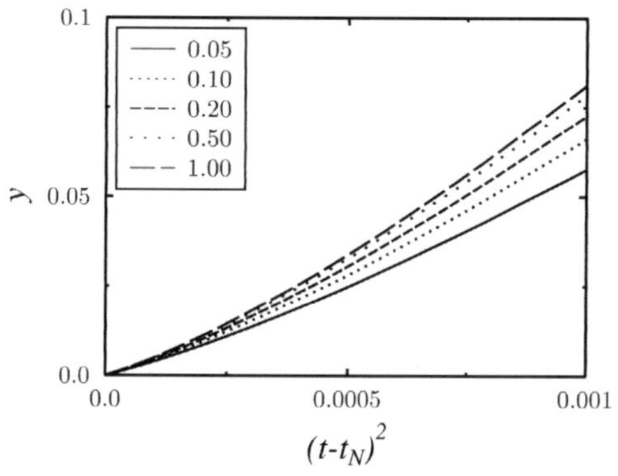

Figure 4. The temperature dependence of $y(t)$ for AF with $t_N = 0.01$.

3. Temperature dependence of thermal expansions

We employ the following free energy [3] consistent with the spin amplitude conservation to discuss the temperature dependence of thermal expansions, given for paramagnetic cases by

$$F_m(y,T,V) = \frac{3}{\pi}\sum_q \int_0^\infty \mathrm{d}\nu \left[\frac{\nu}{2} + T\ln(1-\mathrm{e}^{-\nu/T})\right]\frac{\Gamma_{\mathbf{Q}+\mathbf{q}}}{\nu^2+\Gamma^2_{\mathbf{Q}+\mathbf{q}}} - N_0 T_A y \left\langle \mathbf{S}^2_{loc}\right\rangle. \tag{17}$$

Because of the second term in the right hand side, its stability condition to the variation of y gives the amplitude conservation (2). The thermal expansion can be derived by introducing the volume dependence of parameters in the free energy. According to Grüneisen's treatment of lattice vibrations, let us introduce the parameters,

$$\gamma_0 = -\frac{\mathrm{d}\ln T_0}{\mathrm{d}\omega}, \quad \gamma_A = -\frac{\mathrm{d}\ln T_A}{\mathrm{d}\omega}, \tag{18}$$

that characterize the volume dependence of spectral parameters, T_0 and T_A [3]. The volume strain is defined by $\omega = \delta V/V$. The thermal volume expansion is then derived from the explicit ω-derivative of F_m under the stability condition of y. The result consists of two terms from different origins. We are particularly concerned in this study with the term from the thermal spin fluctuations for paramagnets near the antiferromagnetic instability given by

$$\omega_t(t) = 3\rho K\gamma_0 T_0 t\int_0^1 \mathrm{d}z\int_0^1 \mathrm{d}x\; u[\ln u - 1/2u - \psi(u)], \quad u = (y+z^2+\varepsilon_a x)/t, \tag{19}$$

where $\psi(u)$ is the digamma function.

At low temperatures, the temperature dependence of the thermal expansion $\omega_t(t)$ can be analytically obtained, showing the following t^2-linear dependence:

$$\begin{aligned}\omega_t(t) &\simeq \rho K T_0 \frac{\gamma_0 t^2}{4\varepsilon_a}\left(\ln\frac{1+y_0+\varepsilon_a}{1+y_0} + 2\sqrt{y_0+\varepsilon_a}\tan^{-1}\frac{1}{\sqrt{y_0+\varepsilon_a}} - 2\sqrt{y_0}\tan^{-1}\frac{1}{\sqrt{y_0}}\right), \\ &\simeq \frac{1}{4}\rho K T_0 \gamma_0 t^2\left[1+\frac{\pi}{\sqrt{y_0+\varepsilon_a}+\sqrt{y_0}}\right],\end{aligned} \tag{20}$$

for y_0, $\varepsilon_a \ll 1$ by using the asymptotic expansion of the digamma function:

$$\ln u - \frac{1}{2u} - \psi(u) \simeq \frac{1}{12u^2}, \quad \text{for } u \gg 1.$$

The reciprocal magnetic susceptibility at $T = 0$ is denoted by y_0 as before. Its t^2-linear coefficient is therefore enhanced proportional to $\varepsilon_a^{-1/2}$ for $y_0 \ll \varepsilon_a$ and $y_0^{-1/2}$ for $\varepsilon_a \ll y_0$.

Figure 5 shows the temperature dependence of the thermal expansion $\omega_t(t)/3\rho K\gamma_0$ against t^2 for $y_0 = 0.01$ and various values of ε_a. Since $y_0 \ll \varepsilon_a$ is satisfied, slopes become steeper as ε_a decreases. In Figure 6, the results with fixed $\varepsilon_a = 0.1$ are shown for y_0 =0.01, 0.05, and 0.1. Slopes of curves also become steeper as decreasing y_0 in agreement with the second line of (20).

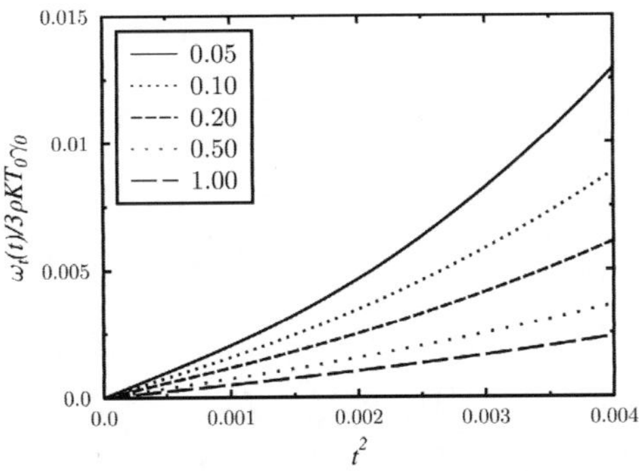

Figure 5. The thermal expansion as a function of t^2 for fixed $y_0 = 0.01$ and ε_a from 0.05 to 1.0 from the top.

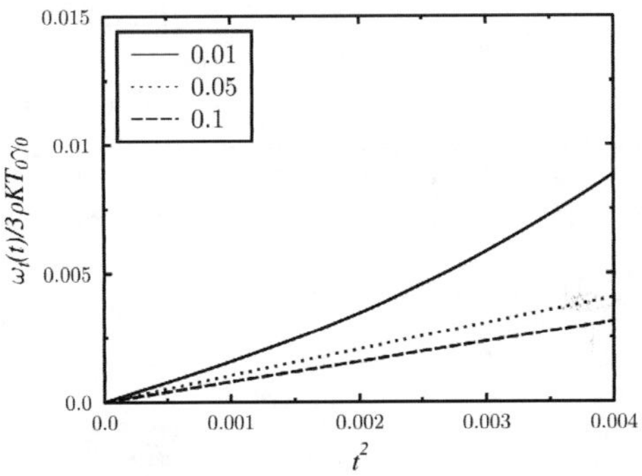

Figure 6. The t^2 linear behavior of $\omega_t(t)$ for y_0 =0.01, 0.05, and 0.1 with fixed $\varepsilon_a = 0.1$.

4. Summary

We have studied magnetic properties and thermal expansion of the quasi one-dimensional itinerant para- and antiferromagnets close to the magnetic instability point. By introducing the anisotropic parameter that controls the relative spectral dispersions along q_z-axis and in q_x, q_y-plane, we can continuously change the spacial dimension from 3-D to 1-D. We have found that magnetic properties especially around the critical point are sensitive to the effect of the change of spacial dimension as summarized below.

It is well-known that the relation, $\sigma_s^2 \propto t_N^{3/2}$, is satisfied for 3-D weak itinerant electron antiferromagnets. We found that it changes into the relation $\sigma_s^2 \propto t_N$ in the 1-D limit. The inverse staggered magnetic susceptibility shows $(t - t_N)^2$-linear dependence around the critical point because of the critical spin fluctuations. The critical region of this behavior is found to increase by lowering the space dimension, for the coefficient of this term decreases towards this limit. The t^2-linear coefficients of the magneto-thermal expansion also show different enhancement around the magnetic instability point depending on the relative magnitude of y_0 and ε_a. Although it is difficult to find one-dimensional antiferromagnets, quasi one-dimensional compounds are in comparison easily available. The present results seem to suggest that quasi-low dimensional itinerant magnets in the extreme anisotropic limit are very interesting in our understanding the quantum critical behaviors from the observed crossovers by changing external parameters, i.e., the temperature, the magnetic field, the volume of crystals, and so on.

Acknowledgments

The authors thank the Yukawa Institute for Theoretical Physics at Kyoto University, where this work was initiated during the YITP-W-10-12 on "International and Interdisciplinary Workshop on Novel Phenomena in Integrated Complex Sciences: from Non-living to Living Systems". This workshop was supported in part by the Grant-in-Aid for the Global COE Program "The Next Generation of Physics, Spun from Universality and Emergence" from the Ministry of Education, Culture, Sports, Science and Technology (MEXT) of Japan. This work was supported by the Kinki University Technical College Research Fund and a JSPS Grant-in-Aid for Science Research (C) 21540341.

References

[1] T. Moriya and K. Usami 1980 *Solid State Commun.* **34** 95
[2] E. P. Wohlfarth 1977 *Physica B* **91** 305
[3] Y. Takahashi and H. Nakano 2006 *J. Phys.: Condens. Matter* **18** 521
[4] Y. Takahashi, H. Nakano, and R. Konno 2007 *J. Magn. Magn. Mater.* **310** 1056
[5] R. Konno, Y. Takahashi, and H. Nakano 2007 *J. Appl. Phys.* **101** 09G517
[6] Y. Takahashi 2001 *J. Phys.: Condens. Matter* **13** 6323
[7] R. Konno 2003 *Physica B* **329-333** 1288

Volume effects and pressure effects on ferromagnetic superconductors

Rikio Konno, Nobukuni Hatayama

Kinki University Technical College, 2800 Arima-cho, Kumano-shi, Mie 519-4395, Japan

E-mail: r-konno@ktc.ac.jp

Abstract. Thermal expansion of ferromagnetic superconductors below the superconducting transition temperature $T_{sc\uparrow}$ of a majority spin conduction band is re-examined. In the previous study [*J.Phys.Conf.Ser.* **200** 012056 (2010)] the volume differential of the kinetic energy of conduction electrons is constant. In this study, the volume differential of the kinetic energy of conduction electrons is inconstant, however. The superconducting gaps used in this study are like those of the thin film of A2 phase in liquid ^{3}He. We find that the thermal expansion coefficient has the divergence at the superconducting transition temperatures. Thermodynamic Grüneisen's relation is satisfied. The pressure effects on the superconductors is also investigated. As an example, the pressure differential of the superconducting transition temperatures and that of the Curie temperature are studied. We find the analytical expression of the pressure differential of the superconducting transition temperatures and that of the Curie temperature.

1. Introduction

The ferromagnetic superconductors [1, 2] has attracted many researchers again since the ferromagnetic superconductors in UGe_2 [3], UCoGe [4] and URhGe [5] was discovered. Recently, Hatayama and Konno [6] investigated the volume effects on ferromagnetic superconductors based on the free energy derived by Linder and Sudbo [7] by applying Takahashi's method [8].

The first purpose of this study is stated. It is assumed that the volume differential of the kinetic energy of conduction electrons is constant in the previous study [6]. The volume differential of the kinetic energy of conduction electrons is usually inconstant. In this study, thermal expansion behaviour is re-examined by making the volume differential of the kinetic energy of conduction electrons inconstant. It is assumed that the Curie temperature T_C is much larger than the superconducting transition temperatures $T_{sc\uparrow}$ of a majority spin band and that the Fermi temperature T_F is much higher than the Curie temperature T_C.

Next, we mention the antecedent studies of the pressure effects on magnetism and superconductivity chronologically. For example, Pfleiderer et al. studied critical behaviour at the magnetic transition as a function of hydrostatic pressure phenomenologically [9]. Aso et al. obtained the pressure dependence of Stoner gap based on the Stoner model from the neutron intensities [10]. Recently, Shopova and Uzunov investigated the pressure dependence of the superconducting transition temperature and that of the Curie temperature in ferromagnetic superconductors phenomenologically [11]. Shopova and Uzunov used the free energy of the Landau expansion about the superconducting order parameters and the magnetisation. Huang et al. reported the pressure differential of the superconducting transition temperature in $FeSe_{1-x}Te_x$ experimentally [12].

The second purpose of this study is expressed. The pressure effects on ferromagnetic superconductors based on the microscopic model have been unresolved yet. The pressure differential of the superconducting transition temperatures and that of the Curie temperature derived from the single band model by Linder and Subo [7] is analysed in the context of the mean-field theory. We will obtain the analytical expression of the pressure differential of the superconducting transition temperatures and that of the Curie temperature.

This paper is organised as follows. In the next section thermal expansion of ferromagnetic superconductors will be derived. In section 3, the numerical results will be provided. In section 4, the pressure differential of the superconducting transition temperatures and that of the Curie temperature will be derived. Section 5 will be devoted to conclusions.

This work was presented at International and Interdisciplinary Workshop on Novel Phenomena in Integrated Complex Sciences: from Non-living to Living Systems. This workshop was supported in part by the Grant-in-Aid for the Global COE Program "The Next Generation of Physics, Spun from Universality and Emergence" from the Ministry of Education, Culture, Sports, Science and Technology (MEXT) of Japan.

2. The derivation of thermal expansion of ferromagnetic superconductors

Thermal expansion of ferromagnetic superconductors is derived in this section. We begin with the following free energy [6, 7]:

$$F/N = F_0/N + F_T/N \tag{1}$$

with

$$F_0/N = \frac{IM^2}{2} + \sum_\sigma \frac{\Delta_{\sigma,0}^2}{2g} - \sum_\sigma \int_0^{E_F} d\epsilon N(\epsilon) \frac{\sqrt{(\epsilon - \sigma IM - E_F)^2 + \Delta_{\sigma,0}^2}}{2}, \tag{2}$$

$$F_T/N = T \sum_\sigma \int_0^\infty d\epsilon N(\epsilon) \ln(1 + e^{-\sqrt{(\epsilon - \sigma IM - E_F)^2 + \Delta_{\sigma,0}^2/T}}), \tag{3}$$

$$\sigma = 1(\uparrow)\text{or} - 1(\downarrow), \tag{4}$$

$$\Delta_{k\sigma\sigma} = \frac{\Delta_{\sigma,0}}{\sqrt{3/8\pi}} Y_{l=1}^{\sigma}(\theta, \phi) \tag{5}$$

where $N(\epsilon)$ is the density of states. g, M, IM are the effective attractive pairing coupling, the magnetisation and the magnetic exchange energy, respectively. $Y_{l=1}^{\sigma}(\theta, \phi)$ is the spherical harmonics. $\Delta_{k\sigma\sigma}$ is the superconducting gap. We shall consider $\sin\phi = 1$ similar to the A2 phase in liquid ^{3}He. The superconducting order parameters at T=0K are obtained

$$\Delta_{\sigma,0}(0) = 2E_0 e^{-1/c\sqrt{1+\sigma\tilde{M}}} \tag{6}$$

where $\tilde{M} = IM/E_F$ and E_F is the Fermi energy. E_0 is the cutoff energy. E_0/E_F is set to 0.01. The weak-coupling constant $c = gN(0)/2$ is set to 0.2. The temperature dependence of the superconducting order parameters is as follows:

$$\Delta_{\sigma,0}(T) = \Delta_{\sigma,0}(0) \tanh(1.74\sqrt{T_{sc\sigma}/T - 1}) \tag{7}$$

where $T_{sc\sigma}$ is the superconducting transition temperature of the spin σ band. Thermal expansion ω is given by $\omega = -K\frac{\partial F}{\partial V}$. K is the compressibility. By our paying attention to $\frac{\partial \epsilon}{\partial V} = \frac{\partial \ln t}{\partial V}\epsilon$ where t is a transfer integral of electrons, from Eq.(1), thermal expansion of ferromagnetic superconductors is derived as follows:

$$\omega/(NE_F) = \omega_0/(NE_F) + \omega_T/(NE_F) \tag{8}$$

with

$$\omega_0/(NE_F) = -K[\frac{1}{2}\frac{E_F}{I}(\frac{\partial \ln I}{\partial V})\tilde{M}^2 + \frac{1}{2}E_F\sum_\sigma \frac{\partial}{\partial V}(\frac{1}{g})\tilde{\Delta}^2_{\sigma,0}$$
$$-N(0)E_F[\frac{\partial \ln N(0)}{\partial V}\int_0^1 dx \sum_\sigma \frac{\sqrt{(x-\sigma\tilde{M}-1)^2+\tilde{\Delta}^2_{\sigma,0}}}{2}$$
$$+\frac{1}{2}\sum_\sigma A_\sigma(\sqrt{(-\sigma\tilde{M})^2} - \sqrt{(-\sigma\tilde{M}-1)^2+\tilde{\Delta}^2_{\sigma,0}})$$
$$+\sum_\sigma \frac{\partial \ln t}{\partial V}\int_0^1 dx x \frac{x-1-\sigma\tilde{M}}{2\sqrt{(x-\sigma\tilde{M}-1)^2+\tilde{\Delta}^2_{\sigma,0}}}]], \quad (9)$$

$$A_\sigma = -\sigma\frac{\partial \ln I}{\partial V}\tilde{M} - \frac{\partial \ln E_F}{\partial V} \quad (10)$$

where $x = \epsilon/E_F$ and $\tilde{\Delta}_{\sigma,0} = \Delta_{\sigma,0}/E_F$. $N(0)$ is the density of states at the Fermi energy. N is the number of the magnetic atoms. The derivatives of $\Delta_{\sigma,0}$ and $\tilde{M}$ do not appear in Eqs.(8), (9), and (10) because of the minimal conditions $\frac{\partial F}{\partial \Delta_{\sigma,0}} = 0$ and $\frac{\partial F}{\partial M} = 0$. The thermal expansion ω_T originated from the explicitly thermal part of the free energy F_T is negligible because the temperature is very low compared with the Fermi temperature T_F. The corresponding thermal expansion coefficient is given by

$$\alpha = \frac{\partial \omega}{\partial T}. \quad (11)$$

In the next section, the temperature dependence of the thermal expansion coefficient of ferromagnetic superconductors will be provided numerically.

3. Results

Numerical results of the temperature dependence of the thermal expansion coefficient of ferromagnetic superconductors is given in this section. The temperature dependence of the thermal expansion coefficient of ferromagnetic superconductors is investigated with Eqs.(6), (7), (8), (9), (10), and (11). Fig.1 shows the temperature dependence of the thermal expansion coefficient. Fig.2 shows the temperature dependence of the thermal expanision coefficient in the vicinity of $T_{sc\downarrow}$. The divergence of the thermal expansion coefficient appears at $T_{sc\sigma}$. The temperature dependence originates from the superconducting order parameters $\Delta_{\sigma,0}(T)$ because the magnetisation at low temperatures is constant and because $T_{sc\sigma}$ is much lower than the Curie temperature. These results show the similar behaviour to the previous study [6]. Moreover, the thermodynamic Grüneisen's relation between the temperature dependence of the thermal expansion and that of the magnetic specific heat is satisfied because we use the same free energy as the free energy when the expression of the specific heat is derived. We will derive the pressure differential of the superconducting transition temperatures $T_{sc\sigma}$ and that of the Curie temperature T_C in the next section.

4. The pressure differential of the superconducting transition temperatures and that of the Curie temperature

In this section, the pressure differential of the superconducting transition temperatures $T_{sc\sigma}$ and that of the Curie temperature T_C. The superconducting transition temperatures are given by [7]

$$T_{sc\sigma} = 1.13E_0 \exp(-1/c\sqrt{1+\sigma\tilde{M}(T_{sc\sigma})}). \quad (12)$$

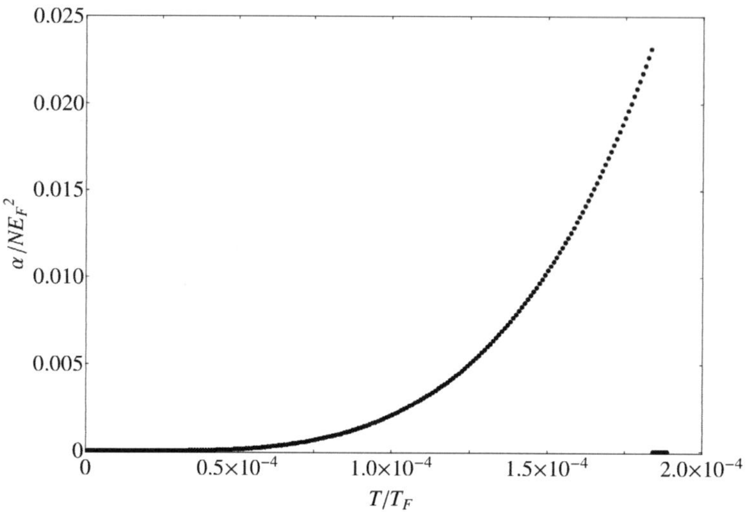

Figure 1. The reduced temperature T/T_F dependence of the thermal expansion coefficient of ferromagnetic superconductors when $E_F/I = 0.1$, $\frac{\partial \ln I}{\partial V} = 0.1$, $E_F \frac{\partial}{\partial V}(\frac{1}{g}) = 0.1$, $N(0)E_F \frac{\partial t}{\partial V} = 0.1$, $N(0)E_F = 0.1$, and $N(0)E_F \frac{\partial \ln N(0)}{\partial V} = 0.1$.

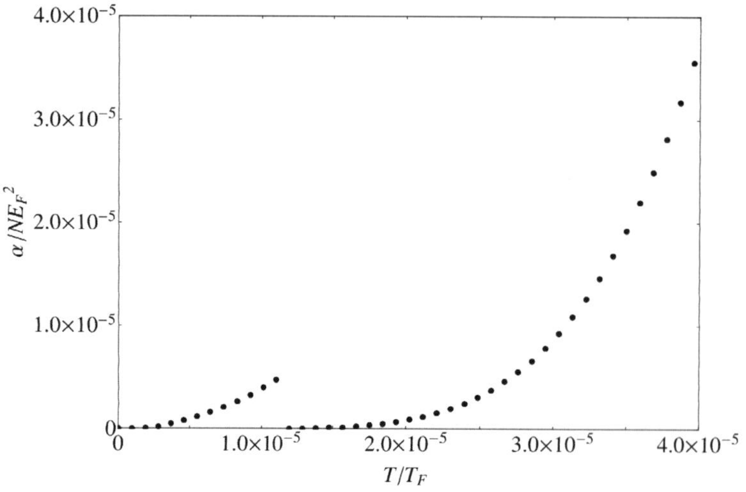

Figure 2. The reduced temperature dependence of the thermal expansion coefficient at very low temperatures when $E_F/I = 0.1$, $\frac{\partial \ln I}{\partial V} = 0.1$, $E_F \frac{\partial}{\partial V}(\frac{1}{g}) = 0.1$, $N(0)E_F \frac{\partial t}{\partial V} = 0.1$, $N(0)E_F = 0.1$, and $N(0)E_F \frac{\partial \ln N(0)}{\partial V} = 0.1$.

This expression is differentiated about the pressure P. We get the pressure differential of the superconducting transition temperature as follows:

$$\frac{\partial T_{sc\sigma}}{\partial P} = T_{sc\sigma}[\frac{\partial}{\partial P}(\ln 1.13E_0) + \frac{1}{c^2\sqrt{1+\sigma\tilde{M}(T_{sc\sigma})}}\frac{\partial c}{\partial P} + \frac{1}{2c(1+\sigma\tilde{M}(T_{sc\sigma}))^{3/2}}\frac{\partial(\sigma\tilde{M}(T_{sc\sigma}))}{\partial P}]. \quad (13)$$

The pressure differential of the cutoff energy E_0, that of the magnetisation, and that of both of the weak coupling constant and density of states will be able to be estimated if this pressure differential of the superconducting transition temperatures and the magnetisation is compared with the experimental data.

We proceed to the pressure differential of the Curie temperature T_C. The magnetisation is determined by the following equation [7]

$$M = -\frac{1}{2}\sum_{\sigma}\sigma\int_0^{\infty} d\epsilon N(0)\tanh[\frac{(\epsilon - E_F - \sigma IM)}{2T}]. \quad (14)$$

In the right hand-side, we perform the integration about ϵ. We get the equation of the magnetisation

$$M = IN(0)M + TN(0)\ln\frac{\cosh(\frac{E_F+IM}{2T})}{\cosh(\frac{E_F-IM}{2T})}. \quad (15)$$

In $T \to T_C$, this equation is expanded around the small M. After that, M is equal to zero at $T = T_C$. We obtain the Curie temperature

$$T_C = \frac{T_F}{\ln(2IN(0)-1)}. \quad (16)$$

If $1 < IN(0) < \frac{e+1}{2}$, the Curie temperature T_C is higher than the Fermi temperature T_F. If $\frac{e+1}{2} < IN(0)$, the Curie temperature T_C is lower than the Fermi temperature T_F. From Eq. (16) we differentiate the Curie temperature T_C about the pressure P. We obtain the pressure differential of the Curie temperature T_C

$$\frac{\partial T_C}{\partial P} = \frac{\frac{\partial T_F}{\partial P}}{\ln(2IN(0)-1)} - \frac{2T_F\frac{\partial(IN(0))}{\partial P}}{(2IN(0)-1)(\ln(2IN(0)-1))^2}. \quad (17)$$

If we analyse the pressure differential of T_C in experimental data by this theory, the pressure differential of T_F and that of $IN(0)$ will be able to be estimated.

5. Conclusions

The thermal expansion on ferromagnetic superconductors by making the volume differential of the kinetic energy of electrons inconstant have been investigated. The superconducting gaps similar to those of the thin film of the A2 phase in liquid ^{3}He are assumed. We find that the divergence of the thermal expansion coefficient exists at the superconducting transition temperatures like the previous study [6]. The thermodynamic Grüneisen's relation between the temperature dependence of the thermal expansion coefficient and that of the magnetic specific heat is satisfied.

The expression of the Curie temperature is derived. The Curie temperature depends on the inverse of logarithm of the electron-electron interaction.

The pressure effects on ferromagnetic superconductors have been analyzed. Both analytical expression of the pressure differential of the superconducting transition temperatures and that of the Curie temperature have obtained. If the expression of the pressure differential of the superconducting transition temperatures and that of the Curie temperature are compared with the experimental data, the pressure differential of the Fermi temperature, that of the weak coupling constant and that of the density of states will be determined.

5.1. Acknowledgments

We would like to thank Y. Takahashi, O. Stockert, M. Kanno , and M. Nakamori for stimulating conversations. One of the authors (R. K.) would like to also thank K. Grube, H. v. Löhneysen, F. Steglich, A. de Visser, M. B. Maple and K. Ishida for stimulating conversations and T. Mamedov and M Maksymenko for sending him preprints prior to the publications. This work is supported by the Kinki University Technical College Research Fund. The authors thank the Yukawa Institute for Theoretical Physics at Kyoto University, where this work was completed during the YITP-W-10-12 on "International and Interdisciplinary Workshop on Novel Phenomena in Integrated Complex Sciences: from Non-living to Living Systems".

References

[1] Tachiki M and Maekawa S 1984 *Phys. Rev.* B **29** 2497
[2] Matsumoto H, Teshima R, Umezawa H, Tachiki M 1983 *Phys. Rev.* B **27** 158 and references therein.
[3] Saxena S S, Agarwal P, Ahilan K, Grosch F M, Haselwimmer R K W, Steiner M J, Pugh E, Walker I R, Julian S R, Monthoux P, Lonzarich G G, Huxley A, Sheikin I, Braithwaite D, and Flouque J 2000 *Nature* **406** 587
[4] Huy N T, Gasparini A, Nijs D E de, Huang Y K, Klaasse J C P, Gortenmulder T, Visser A de, Hamann A, Görlach T, and Löhneysen v H 2007 *Phys. Rev. Lett* **99** 067006, and references therein.
[5] Levy F, Sheikin I, Grenier B, Marcenat C, and Huxley A 2009 *J. Phys.:Condens. Matter* **21** 164211, and references therein.
[6] Hatayama N and Konno R 2010 *J. Phys. Conf. Ser.* **200** 012056
[7] Linder J and Sudbo 2007 *Phys. Rev.* B **76** 054511
[8] Takahashi Y and Nakano H 2006 *J. Phys. Condens. Matter* **18**, 521
[9] Pfleiderer C, MeMullan G J, Lonzarich G G 1994 *Physica* **B 199-200**, 634
[10] Aso N, Motoyama G, Uwatoko Y, Ban S, Nakamura S, Nishioka T, Homma Y, Shiokawa Y, Hirota K, and N. K. Sato 2006 *Phys. Rev.* B **29** 054512
[11] Shopova D V and Uzunov D I 2009 *Phys. Rev.* B **79** 064501
[12] Huang C L, Chou C C, Tseng K F, Huang Y L, Hsu F C, Yeh K W, Wu M K, Yang H D 2009 *J. Phys. Soc. Jpn.* **78** 084710

IIW-NPICS 2010 IOP Publishing
Journal of Physics: Conference Series **344** (2012) 012017 doi:10.1088/1742-6596/344/1/012017

Magnetic susceptibility of η-carbide-type molybdenum and tungsten carbides and nitrides

T Waki, S Terazawa, Y Umemoto, Y Tabata, Y Murase[1], M Kato[1], K Hirota[1], H Nakamura

Department of Materials Science and Engineering, Kyoto University, Kyoto 606-8501, JAPAN
[1] Department of Molecular Science and Technology, Doshisha University, Kyotanabe, Kyoto 610-0394 JAPAN

E-mail: waki.takeshi.5c@kyoto-u.ac.jp

Abstract. We synthesized a series of molybdenum and tungsten η-carbide-type compounds and measured their magnetic susceptibility χ. Although the enhancement of χ was observed in several compounds, apparent magnetic order was found only in Fe_3W_3N (ferromagnetic at 110 K). The broad peak in χ of Fe_6W_6C is possibly due to antiferromagnetic order. Itinerant electron metamagnetic transitions were observed in Fe_3W_3C, Fe_6W_6C, Fe_6W_6N and Co_6W_6C. The magnetic enhancement depends on the $3d$ element in the order Fe > Co > Ni corresponding to DOS at theFermi energy in η-12 systems, whereas no clear trend has been observed in η-6 systems.

1. Introduction

The η-carbide-type compounds with the cubic Fe_3W_3C-type crystal structure (space group $Fd\bar{3}m$), are known as refractory and hard materials, and recently as potential catalysts. The general formula is $M_6^f T_4^e T_2^d X_m^c X_n^a$, where M = Nb, Mo, W, Ta, $\cdots$, T = Cr, Mn, Fe, Co, Ni, $\cdots$, X = C, N, O, $m = 0, 2$, $n = 0, 1$ and the superscripts of the metal sites represent Wyckoff positons. In the case of $m = 2$ and $n = 0$, namely T_3M_3X, the structure is referred to as η-6 (the number denotes the ratio of metal atoms against the X atom). Similarly $m = 0$ and $n = 1$, T_6M_6X is called η-12. In both structures, the three dimensional T metal network consists of the 16d pyrochlore lattice nested by 32e tetrahedrons called a *stella quadrangula* [1], which is of interest in the viewpoint of geometrical frustration. As the T-T distance in η-carbide-type compounds is short of 2.5 $\sim$ 2.6 Å, the itinerant electron magnetism is expected. Although many studies were devoted on these materials, few quantum properties have been reported so far. Recently we have found a ferromagnetic quantum critical behavior in Fe_3Mo_3N [2] and also field and impurity induced ferromagnetic orders in the compound [3]. As the ferromagnetic orders are triggered by slight perturbations, the absence of magnetic long-range order in pure Fe_3Mo_3N could be due to the geometrical frustration in the stella quadrangula lattice of the Fe sublattice [3]. We also found an itinerant electron metamagnetism (IEM) in Co_3Mo_3C [4]. Therefore, a rich variety of magnetism is expected in other η-carbide-type compounds . In this report we show temperature dependences of χ measured for several molybdenum and tungsten η-6 and η-12 compounds.

2. Experiments

We synthesized polycrystalline samples of Mn_3Mo_3C, Fe_3W_3N, Fe_3W_3C, Fe_3Mo_3C, Fe_6W_6C, Fe_6W_6N, Co_6W_6C, Co_6Mo_6N, Ni_6Mo_6C, and Ni_6W_6C. Except Fe_6W_6N, the synthesis has been reported in the literature together with the crystallographic information [5, 6, 7, 8, 9, 10, 11, 12, 13]. Most carbides were obtained using a standard solid state reaction from powdered elements in evacuated silica tubes. Fe_3Mo_3C was obtained by two methods; one is the solid state reaction (#1) and the other is a topotactic method from the nitride (#2) [12]. On the other hand, the nitrides were obtained using sprcific methods; Co_6Mo_6N was obtained via a topotactic route from Co_3Mo_3N described in the reference [5], Fe_3W_3N was successfully synthesized using a hot isostatic pressing method starting from metal elements in N_2 high-pressure gas. Fe_6W_6N was also obtained by the same method under different pressures and temperatures. The synthesized samples were checked by the powder XRD method. Except Mn_3Mo_3C and Fe_6W_6N, we obtained single phases. As impurities, MoC and Fe_3W_3N were detected in Mn_3Mo_3C and Fe_6W_6N, respectively. The temperature dependence of magnetic susceptibility χ was measured in the temperature range from 1.8 to 300 K using a SQUID magnetometer, MPMS equipped in the LTM center, Kyoto University. High field magnetization measurements were performed using a pulse magnet equipped in ISSP, the University of Tokyo, up to 50 T at 4.2 K.

3. Results and Discussion

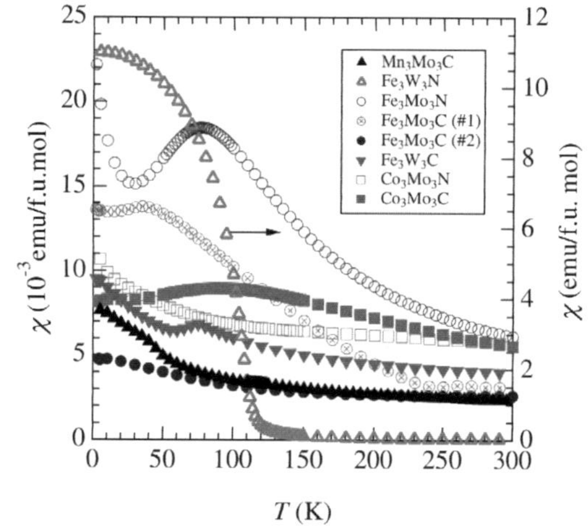

Figure 1. χ for η-6.

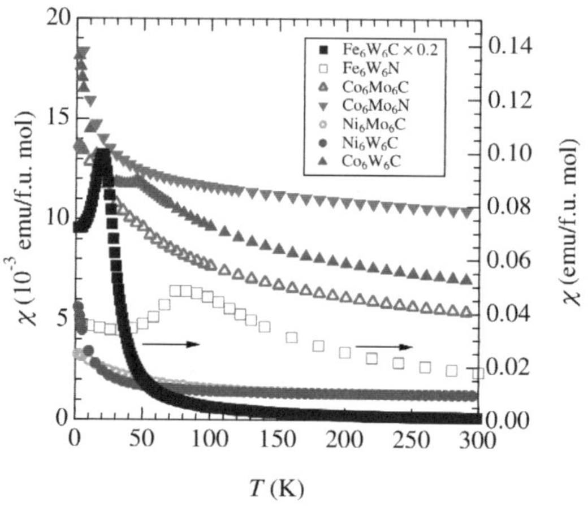

Figure 2. χ for η-12.

Temperature dependences of χ for η-6 and η-12 compounds are summarized in Figs. 1 and 2, respectively. The data of Fe_3Mo_3N, Co_3Mo_3C and Co_3Mo_3N are taken from our previous reports [2, 4]. Most of the systems show Curie-Weiss (CW) like temperature dependences at high temperatures, suggesting magnetic enhancement in these compounds. The CW susceptibilities were fitted to a modified CW function $\chi = C/(T-\theta)+\chi_0$, where $C = Np_{\rm eff}^2\mu_{\rm B}^2/3k_{\rm B}$ is the Curie constant, N the number of magnetic atoms, $p_{\rm eff}$ the effective moment, $\mu_{\rm B}$ the Bohr magneton, $k_{\rm B}$ the Boltzmann constant, θ the Weiss temperature, and χ_0 a temperature independent term of the susceptibility. The estimated $p_{\rm eff}$ and θ are summarized in Table 1.

Positive θ values often correspond the Curie temperature of the system, although negative θ values in the itinerant electron system do not necessarily mean antiferromagnetic interaction. As expected from the θ value, Fe_3W_3N undergoes ferromagnetic order at 110 K. The ratio of $p_{\rm eff}/p_{\rm s} = 4.8$, estimated by using the saturation moment at 5 K $p_{\rm s} = 0.76\mu_{\rm B}/{\rm Fe}$, is larger than unity, which is one of characteristics for the weak itinerant electron ferromagnet. A positive θ value was also obtained for Fe_6W_6C. However, a pronounced peak in χ(T) instead of ferromagnetic order has been observed. The magnetization at 2 K shows a sudden increase at 0.55 T suggesting a metamagnetic transition (MT). This transition is possibly of spin flip in the antiferromagnet because the critical field $H_C(T)$ decreases with increasing temperature, contrary to the increase in $H_C(T)$ generally observed for IEM of exchange-enhanced nearly ferromagnetic metals. Interestingly, Fe_3Mo_3N, with almost zero θ, shows a ferromagnetic non-Fermi liquid behavior at low temperature [2]. In that sense, Fe_3W_3C deserves a study for the quantum critical magnetism because it shows almost zero θ as well.

We observed $\chi(T)$ maximum behaviors in several systems; Fe_3Mo_3N, Fe_3W_3C, Co_3Mo_3C, Fe_3Mo_3C(#1), Co_6W_6C and Fe_6W_6N. Because the $\chi(T)$ maximum has been commonly observed in exchange-enhanced itinerant electron metals showing IEM, we performed high field magnetization measurements on these materials. We have already reported the IEMs in Fe_3Mo_3N [3] and Co_3Mo_3C [4]. By pulse magnetization measurements, we observed IEM at 27.8, 27 and 26.2 T in Fe_3W_3C, Fe_6W_6N and Co_6W_6C, respectively [14, 15]. Note that the result of Fe_6W_6N is only preliminary because our Fe_6W_6N sample contains extra phases. Unfortunately we have not established synthetic conditions to get a single phase of Fe_6W_6N.

No apparent anomalies were observed in $\chi(T)$ of Mn_3Mo_3C and Co_6Mo_6C, although CW-type temperature dependences were seen. Weak temperature dependences of χ were observed for Fe_3Mo_3C(#2), Co_3Mo_3N and Co_6Mo_6N. As the magnitudes are relatively large, there should be exchange enhancements. $\chi(T)$s of Ni_6Mo_6C and Ni_6W_6C remain small suggesting simple Pauli paramagnetism in these materials.

In Fe_3Mo_3C, the difference in magnetism was observed depending on synthetic routes. $\chi(T)$ of (#1) shows Pauli paramagnetism at high temperature and a gradual increase below 250 K being different from a CW behavior. On the contrary, $\chi(T)$ of (#2) is almost temperature independent. The difference may be due to the presence of excess Fe or Mo atoms occupying the counter metallic site which depends the manufacturing method [6, 12]. We note that no metamagnetic transition was observed in Fe_3Mo_3C(#1) up to 45 T, although a $\chi(T)$ maximum was observed at 35 K. The origin of the maximum might be different from those for other materials.

Among η-12 systems, a clear correlation between the enhancement of χ and the T element can be seen; with increasing the number of electrons as changing T from Fe to Ni, the magnetic enhancement deceases. For η-12 systems, only few results of band calculations are available [16]. The bands of Fe_6W_6C and Co_6W_6C consist of pronounced $3d$ bands and widely spread $5d$ W bands whose centers of gravity locate lower and higher than the Fermi energy, respectively. When we assume the rigid band model, the contribution of $5d$ bands at the Fermi energy increases with increasing the number of electrons, resulting in a magnetically less enhanced state. On the other hand, we have not observed an appreciable trend in the magnetism among η-6 systems, although the magnetism depends sensitively on the elements at the M and T sites. Band structures of Fe_3W_3C and Co_3W_3C (this compound is metastable at room temperature [7]), reported in the literature, do not show marked difference from those of η-12.

4. Conclusion

We measured temperature dependences of the susceptibility for several molybdenum and tungsten η-carbide-type compounds and found some magnetically enhanced materials. An apparent magnetic long-range order was observed only in Fe_3W_3C (ferromagnetic long range order at 110 K). Fe_6W_6C is possibly antiferromagnetic. $\chi(T)$ maximum behaviors, which remind us IEM, were observed in several systems. We have found IEM in Fe_3W_3C, Fe_6W_6N and Co_6W_6C.

Table 1. Magnetic properties of η-carbide-type compounds

materials	$p_{\rm eff}$ (μ_B/3d-atom)	θ (K)	memo
Mn_3Mo_3C	0.98	-65	CW
Fe_3Mo_3C (#1)	-	-	$\chi(T)$ max @ 35 K
Fe_3Mo_3C (#2)	-	-	Pauli paramagnetic
Fe_3Mo_3N	2.14	2	$\chi(T)$ max @ 78 K, IEM @ 13.9 T
Fe_3W_3C	0.85	-1	$\chi(T)$ max @ 75 K, IEM @ 27.8 T
Fe_3W_3N	3.39	110	F @ 110 K
Co_3Mo_3C	2.50	-124	$\chi(T)$ max @ 90 K, IEM @ 37 T
Co_3Mo_3N	-	-	Pauli paramagnetic
Fe_6W_6C	1.50	28	$\chi(T)$ max @ 20 K, MT @ 0.55 T
Fe_6W_6N	2.74	-21	$\chi(T)$ max @ 80 K, IEM @ 27 T
Co_6W_6C	1.13	-93	$\chi(T)$ max @ 46 K, IEM @ 26.2 T
Co_6Mo_6C	0.99	-73	CW
Co_6Mo_6N	-	-	Pauli paramagnetic
Ni_6Mo_6C	-	-	Pauli paramagnetic
Ni_6W_6C	-	-	Pauli paramagnetic

Acknowledgments

The authors thank A. Kondo, K. Sato and K. Kindo at ISSP, the University of Tokyo for the collaboration in high-field magnetization measurements, and S. Alconchel at Universidad Nacional del Litoral and F. Sapiña at Universitat de València for providing Fe_3Mo_3C (#2) sample. The authors also thank the Yukawa Institute for Theoretical Physics at Kyoto University, where this work was initiated during the YITP-W-10-12 on "International and Interdisciplinary Workshop on Novel Phenomena in Integrated Complex Sciences: from Non-living to Living Systems". This study was supported by a Grant-in-Aid for Scientific Research on Priority Areas "Novel States of Matter Induced by Frustration" (19052003), a Grant-in-Aid for the Global COE Program, "International Centre for Integrated Research and Advanced Education in Materials Science," a Grant-in-Aid for Young Scientists (B) 21760531 from MEXT of Japan, and a grant from Kinki Regional Invention Center (Foundation).

References

[1] Nyman H, Andersson S and O'Keeffe M 1978 *J. Solid State Chem.* **26** 123
[2] Waki T *et al.* 2010 *J. Phys. Soc. Jpn.* **79** 043701
[3] Waki T *et al.* 2011 *EPL* **94** 37004
[4] Waki T *et al.* 2010 *J. Phys. Soc. Jpn.* **79** 093703
[5] Hunter S M, Mckay D, Smith R I, Hargreaves J S J and Gregory D H, 2010 *Chem. Mater.* **22** 2898
[6] Fraker A C and Stadelmaier H H 1969 *Trans. Metal. AIME* **245** 847
[7] Pollock C B and Stadelmaier H H 1970 *Metal. Trans.* **1** 767
[8] Telegus V S, Kuz'ma Y B and Marko M A 1971 *Poroshkovaya Met.* **11** 56
[9] Ettmayer P and Suchentrunk R 1970 *Monatsh. Chem.* **101** 1098
[10] Weil K S and Kumta P N 1997 *Acta. Cryst.* C **53** 1745
[11] Newsam J M, Jacobson A J, McCandlish L E and Polizzotti R S 1988 *J. Solid State Chem.* **75** 296
[12] Alconchel S, Sapiña F and Martínez E 2004 *Dalton Trans.* 2463
[13] Prior T J and Battle P D 2004 *J. Solid State Chem.* **14** 3001
[14] Waki T *et al.* 2011 *J. Phys.: Conf. Ser.* **320** 012069
[15] Waki T *et al.* 2011 *J. Alloys Compd.* **509** 9451
[16] Suetin D V, Shein I R and Ivanovskii A L 2009 *Physica* B **404** 3544

Journal of Physics: Conference Series **344** (2012) 012018 doi:10.1088/1742-6596/344/1/012018

The Relation between the Electric Conductance of Nanostructure Bridge and Friedel Sum Rule

Y. Kotani[1,2], N. Shima[1,3] and K. Makoshi[1,3]

[1]Graduate School of Material Science, University of Hyogo, Hyogo 678-1297, Japan
[2]Synchrotron Radiation Research Center, Japan Atomic Energy Agency, Hyogo 679-5148, Japan
[3]CREST, Japan Science Technology, Saitama 332-0012, Japan

E-mail: rk08j001@stkt.u-hyogo.ac.jp

Abstract. We analyze the electric conductance through nanostructure bridges in terms of phase-shifts, which satisfy the Friedel sum rule. The phase-shifts are given by solving the eigenvalue equation obtained by extending the method applied to a single impurity problem in a metal. The local charge neutrality condition is introduced through the Friedel sum rule. It is analytically shown that the electric conductance can increase as the two electrodes separate with the condition in which the phase-shifts satisfy the Friedel sum rule. The increment of the distance between two electrodes is obtained by gradually separating interatomic distance.

1. Introduction

Recently, the electron transport through a nanostructure bridging two electrodes has been a subject of considerable investigation. In the experiments involving Al and Pt, the increment of the electric conductance has been observed as the two electrodes separate and this phenomena has theoretically investigated.[1,2] Most of theoretical studies have been based on the electronic states calculations. Among them, there has been a theory based on the phase-shifts analysis with the Friedel sum rule (FSR).[3] Datta and Tian have discussed the single orbital case with the inversion symmetry of the whole system based on the phase-shifts analysis.[4] Moreover Makoshi, Shima and Otsuka have extended their study to general case.[5]

In an atomic contact, the charge neutrality is very important to understand the behavior of the electric conductance.[7,8] In the tight-binding approximation, this effect can be self-consistently corrected imposing local charge neutrality.[7,8] It is considered that the charge relates electric conductance, because some experiments is reproduced by imposing the charge neutrality condition on the models. However the electric conductance described by general calculation does not involve the information about local charge.

On the other hand, the electric conductance described by the phase-shift is directly related to charge through the FSR. Moreover the phase-shift analysis can analytically introduce the charge neutrality condition. Consequently, we consider that phase-shift analysis gives more understanding of the electron transport property through a nanostructure bridges.

In this paper, we analyze the electric conductance through nanostructure bridges in terms of phase-shifts and construct the theory directly relating to the charge through the FSR. The phase-shifts are given by solving the eigenvalue equation obtained by extending the method applied

Journal of Physics: Conference Series **344** (2012) 012018 doi:10.1088/1742-6596/344/1/012018

to a single impurity problem in a metal. It is analytically shown that the electric conductances through nanostructure bridges consisting of an atom such as Al and Pt can increase when the distance between two electrodes is larger when the phase-shifts satisfy the FSR. We suppose that increment of the distance between two electrodes is obtained by gradually separating interatomic distance.

This workshop was supported in part by the Grant-in-Aid for the Global COE Program "The Next Generation of Physics, Spun from Universality and Emergence" from the Ministry of Education, Culture, Sports, Science and Technology (MEXT) of Japan.

2. Formalism

The Green's function theory applicable to the nonequilibrium system is employed to formulate the electric conductance.[9–12] We calculate the conductance at 0 bias limit. The relation between Green's functions and phase-shifts are obtained by the extending method applied to a single impurity problem in a metal.[6] We suggest that the electric conductance can be formulated even with electron correlation, as long as the electric conductance is represented by the phase-shifts, although we have not dealt with it yet. The electron correlation effect is neglected as the first step. But a part of the Coulomb interaction is effectively involved in the one particle part of the Hamiltonian. Then the system is spin-independent. Thus, we deal with the system as if it is a spinless fermion system and the final results are doubled. We assume that the system considered here has the time reversal symmetry for simplicity.

The Hamiltonian is

$$H = H_0^{\mathrm{L}} + H_0^{\mathrm{C}} + H_0^{\mathrm{R}} + V, \tag{1}$$

$$H_0^{\mathrm{L}} = \sum_l \epsilon_l a_l^\dagger a_l, \quad H_0^{\mathrm{C}} = \sum_{cc'} h_{cc'} a_c^\dagger a_{c'}, \quad H_0^{\mathrm{R}} = \sum_r \epsilon_r a_r^\dagger a_r, \tag{2}$$

$$V = \sum_{c,l} V_{cl} a_c^\dagger a_l + \sum_{c,r} V_{cr} a_c^\dagger a_r + h.c., \tag{3}$$

where l and r are states, respectively, in the left (L)-electrode, in the right(R)-electrode and c stands for a site number and atomic orbital in the bridging part (C). In this model, the full retarded Green's function $G(\omega)$ is given by

$$G(\omega) = (\omega - H + i0)^{-1}. \tag{4}$$

The expression of the electric conductance at 0 bias limit, i.e., in the linear response regime, $\sigma = (2e^2/h)\bar{\sigma}$, can be obtained[13,14] as

$$\bar{\sigma} = 4\mathrm{Tr}^{\mathrm{C}}[\Delta^{\mathrm{L}}(\epsilon_{\mathrm{F}}) G^{\mathrm{C}}(\epsilon_{\mathrm{F}}) \Delta^{\mathrm{R}}(\epsilon_{\mathrm{F}}) G^{\mathrm{C}\dagger}(\epsilon_{\mathrm{F}})], \tag{5}$$

where ϵ_{F} is the Fermi energy at electrodes. $\mathrm{Tr}^{\mathrm{C}}$is taken over the states in bridging part. Here, $G^{\mathrm{C}}(\omega)$ is the bridging part of the full retarded Green's function $G(\omega)$. This is given by

$$\begin{aligned} G^{\mathrm{C}}(\omega) &= (\omega - H_0^{\mathrm{C}} + i0 - \Sigma)^{-1} & (6)\\ &= (g^{-1} - \Sigma)^{-1}, & (7) \end{aligned}$$

with the retarded free Green's function of the bridging part g. The self-energy part $\Sigma(\omega) = \Lambda(\omega) - i\Delta(\omega)$ is given by

$$\begin{aligned} \Delta_{cc'}(\omega) &= \Delta_{cc'}^{\mathrm{L}}(\omega) + \Delta_{cc'}^{\mathrm{R}}(\omega) & (8)\\ &= \pi \sum_l V_{cl} V_{lc'} \delta(\omega - \epsilon_l) + \pi \sum_r V_{cr} V_{rc'} \delta(\omega - \epsilon_r). & (9) \end{aligned}$$

We define the phase-shifts $\varphi_\alpha(\omega)$ which satisfy the FSR, by the extending Langreth's method[6] given by Makoshi, Shima and Otsuka[5] as

$$G^{\mathrm{C}}(\omega)G^{\mathrm{C}\dagger-1}(\omega)\vec{u}^\alpha = e^{2i\varphi_\alpha(\omega)}\vec{u}^\alpha, \tag{10}$$

where $\vec{u}^\alpha$ is an eigenvector. The phase-shifts take a value from 0 to π. We obtain the excess electron number due to the contact $N_{el}(\epsilon_{\mathrm{F}})$ by the phase-shifts defined in eq.(10) as

$$N_{el}(\epsilon_{\mathrm{F}}) = -\frac{1}{\pi}\int_{-\infty}^{\epsilon_{\mathrm{F}}} \mathrm{d}\omega \mathrm{Im}\mathrm{Tr}[G(\omega) - G_0(\omega)] \tag{11}$$

$$= \frac{2}{\pi}\sum_\alpha \varphi_\alpha(\epsilon_{\mathrm{F}}). \tag{12}$$

Here, $G_0(\omega)$ is defined as

$$G_0(\omega) = \left(\begin{array}{c|c|c} G_0^{\mathrm{L}}(\omega) & 0 & 0 \\ \hline 0 & 0 & 0 \\ \hline 0 & 0 & G_0^{\mathrm{R}}(\omega) \end{array}\right), \tag{13}$$

$$G_0^{\mathrm{L(R)}}(\omega) = (\omega - H_0^{\mathrm{L(R)}} + i0)^{-1}. \tag{14}$$

In our calculation, we introduce the charge neutrality condition through the FSR as

$$Z = N_{el}(\epsilon_{\mathrm{F}}), \tag{15}$$

where Z is the total number of valence electrons.

The simple expression of the electric conductance at 0 bias limit, i.e., in the linear response regime, $\sigma = G_0\bar{\sigma}$, is obtained as follows:[5]

$$\bar{\sigma} = 4\sum_\alpha^M \sin^2\varphi_\alpha[\tilde{P}^\dagger\Delta_{\mathrm{L}}\tilde{P}\tilde{P}^\dagger\Delta_{\mathrm{R}}\tilde{P}] + \sum_{\alpha>\beta}^M \sin^2(\varphi_\alpha - \varphi_\beta)|[\tilde{P}^\dagger(\Delta_{\mathrm{L}} - \Delta_{\mathrm{R}})\tilde{P}|^2_{\alpha\beta}. \tag{16}$$

Here the matrix $\tilde{P}$ $(N \times M)$ is given by

$$\tilde{P} = (\vec{u}^1, \ldots, \vec{u}^M), \tag{17}$$

where N is the total number of sites and atomic orbitals in the bridging part, and M is the number of the phase-shifts contributing to the electric conductance.[5]

3. Examples

We investigate two concrete examples. In these models, we suppose that increment of the distance between two electrodes is obtained by gradually separating interatomic distance. We impose the charge neutrality condition on the phase-shifts through the FSR.

3.1. Single atom bridge with Al-like character

First example is a contact consisting of an atom with *sp*-orbitals such as Al atomic contact shown in Fig.1. We can write H_0^{C} and Δ as

$$H_0^{\mathrm{C}} = \begin{pmatrix} \epsilon_s & 0 & 0 & 0 \\ 0 & \epsilon_z & 0 & 0 \\ 0 & 0 & \epsilon_x & 0 \\ 0 & 0 & 0 & \epsilon_y \end{pmatrix}, \tag{18}$$

$$\Delta = \begin{pmatrix} \Delta_{\rm L}^{sp_z} + \Delta_{\rm R}^{sp_z} & 0 \\ 0 & \Delta_{\rm L}^{xy} + \Delta_{\rm R}^{xy} \end{pmatrix}, \tag{19}$$

$$\Delta_{\rm L}^{sp_z} = \begin{pmatrix} a_1 & a_3 \\ a_3 & a_2 \end{pmatrix}, \quad \Delta_{\rm R}^{sp_z} = \begin{pmatrix} a_1 & -a_3 \\ -a_3 & a_2 \end{pmatrix}, \quad \Delta_{\rm L}^{xy} = \Delta_{\rm R}^{xy} = \begin{pmatrix} b_1 & 0 \\ 0 & b_2 \end{pmatrix}. \tag{20}$$

Now we neglect the effect of Λ for simplicity. In this model, we can show that four phase-shifts φ_s, φ_z, φ_x and φ_y contribute to the conductance as

$$\bar{\sigma} = \sin^2\varphi_s + \sin^2\varphi_z + \sin^2\varphi_x + \sin^2\varphi_y - 2\frac{a_3^2}{a_1 a_2}\sin\varphi_s \sin\varphi_z \cos(\varphi_s - \varphi_z). \tag{21}$$

The electric conductance of eq.(21) takes a value between 0 and $4G_0$. The phase-shifts are defined as

$$\cot\varphi_s = -\frac{\epsilon_{mF} - \epsilon_s}{a_1}, \quad \cot\varphi_z = -\frac{\epsilon_{mF} - \epsilon_z}{a_2}, \quad \cot\varphi_x = -\frac{\epsilon_{mF} - \epsilon_x}{b_1}, \quad \cot\varphi_y = -\frac{\epsilon_{mF} - \epsilon_y}{b_2}. \tag{22}$$

In this model, we impose the charge neutrality condition through the FSR as

$$3 = \frac{2}{\pi}(\varphi_s + \varphi_z + \varphi_x + \varphi_y). \tag{23}$$

Here the change of the site energy from isolated system corresponding to p_x is smaller than the ones of s- and p_z-orbitals and the matrix element corresponding to the π-bond (b_1) is smaller than the ones corresponding to the σ bonds, because the broadening of the wave function of p_x-orbitals in the direction of electrodes is smaller than the ones of s- and p_z-orbitals. The same is true of p_y-orbital. Therefore it is considered that φ_x and φ_y are nearly equal to 0 and the contributions of the phase-shifts corresponding to π-bond is very little under the condition that the phase-shifts satisfy the FSR. Thus φ_s and φ_z mainly contribute to the electric conductance and the electric conductance becomes

$$\bar{\sigma} = \sin^2\varphi_s + \sin^2\varphi_z - 2\frac{a_3^2}{a_1 a_2}\sin\varphi_s \sin\varphi_z \cos(\varphi_s - \varphi_z), \tag{24}$$

and the FSR is represented as

$$3 = \frac{2}{\pi}(\varphi_s + \varphi_z). \tag{25}$$

The relationship between φ_s and φ_z is given by eq.(25) as

$$\cot\varphi_s = \tan\varphi_z, \tag{26}$$

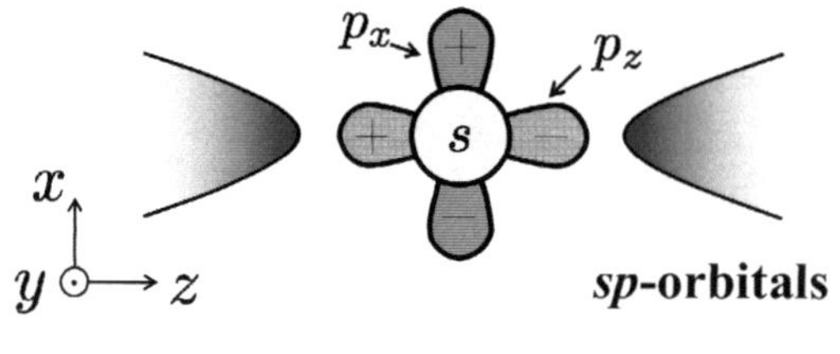

Figure 1. Sketch of system considered here. The bridge is the contact consisting of an atom with *sp*-like character. We consider four orbitals in this model, however we draw only three orbitals s-, p_z- and p_x-orbital in this figure.

Journal of Physics: Conference Series **344** (2012) 012018 doi:10.1088/1742-6596/344/1/012018

or equivalently,

$$\frac{a_1}{\epsilon_{\rm F} - \epsilon_s} = \frac{\epsilon_{\rm F} - \epsilon_z}{a_2}. \tag{27}$$

Finally, the electric conductance is represented as

$$\bar{\sigma} = 1 - \frac{a_3^2}{a_1 a_2} \sin^2 2\varphi_z. \tag{28}$$

We notice that the electric conductance cannot exceed $1G_0$. Here we discuss the increment of the electric conductance when the distance between two electrodes increases. The increment of the distance is obtained by gradually increasing interatomic distance in our work. In other words, this effect is expressed by taking limit value of parameters (a_1 and a_2 tend to 0). In a simple treatment, the phase-shifts in eqs.(22) take values 0 or π when the parameters tend to 0. We notice that the phase-shifts do not satisfy the FSR in this treatment. However when the numerator in the right-hand side takes a value 0 (for example $\epsilon_{\rm F} - \epsilon_z = 0$), the phase-shift is allowed to take a value $\pi/2$ as the limit value. Considering these results, there is the only allowable solution that φ_z tends to $\pi/2$ and φ_s tends to π. Actually in eq. (27), the matrix element corresponding to s-orbital (a_1) converges to 0 faster than that corresponding to p_z-orbital (a_2), because the wave function of p_z-orbital is broader than that of s-orbital. It is shown the solution that the phase-shift takes a value $\pi/2$ is reasonable.

Therefore the phase-shift φ_z tends to $\pi/2$ and the electric conductance approaches to $1G_0$ leading to the increment, when the matrix elements (a_1 and a_2) tend to 0 under the condition that the phase-shifts satisfy the FSR. This phenomenon is interpreted by the FSR and this result is consistent with the experiments.[1,2]

3.2. Single atom bridge with Pt-like character

Second example is a contact consisting of Pt atom with sd-orbital shown in Fig.2. We can write $H_0^{\rm C}$ as

$$H_0^{\rm C} = \begin{pmatrix} \epsilon_s & 0 & 0 & 0 & 0 & 0 \\ 0 & \epsilon_{3z^2-r^2} & 0 & 0 & 0 & 0 \\ 0 & 0 & \epsilon_{yz} & 0 & 0 & 0 \\ 0 & 0 & 0 & \epsilon_{zx} & 0 & 0 \\ 0 & 0 & 0 & 0 & \epsilon_{xy} & 0 \\ 0 & 0 & 0 & 0 & 0 & \epsilon_{x^2-y^2} \end{pmatrix}. \tag{29}$$

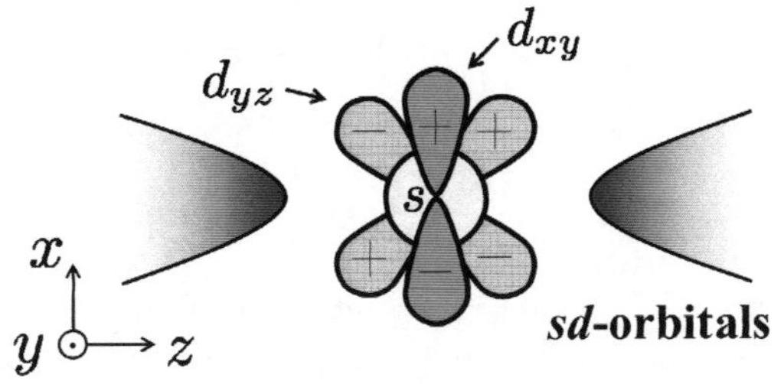

Figure 2. Sketch of system considered here. The bridge is the contact consisting of an atom with sd-like character. We consider the six orbitals in this model, however we draw only three orbitals s-, d_{xy}- and d_{yz}-orbital in this figure.

Then Δ is written as

$$\Delta = \begin{pmatrix} \Delta_\sigma & 0 & 0 \\ 0 & \Delta_\pi & 0 \\ 0 & 0 & \Delta_\delta \end{pmatrix}, \tag{30}$$

$$\Delta_\sigma = \begin{pmatrix} a_1 & a_3 \\ a_3 & a_2 \end{pmatrix}, \quad \Delta_\pi = \begin{pmatrix} b & 0 \\ 0 & b \end{pmatrix}, \quad \Delta_\delta = \begin{pmatrix} c & 0 \\ 0 & c \end{pmatrix}, \tag{31}$$

where Δ_σ, Δ_π and Δ_δ correspond to σ-, π- and δ-bond between electrodes and bridging part respectively. Again we neglect the effect of Λ for simplicity. Because of the symmetry of this model, one can consider that the site energies corresponding to π-bond degenerate (i.e. $\epsilon_{yz}=\epsilon_{zx}=\epsilon_\pi$). Also those corresponding to δ-bond degenerate (i.e. $\epsilon_{xy}=\epsilon_{x^2-y^2}=\epsilon_\delta$). Moreover because of the strength of the interaction between each atomic orbital and electrodes, Fermi energy and the site energies have the relationship as follows:

$$\epsilon_{\mathrm{F}} > \epsilon_\delta > \epsilon_\pi > \epsilon_{3z^2-r^2} > \epsilon_s. \tag{32}$$

We can show that only four phase-shifts φ_s, $\varphi_{3z^2-r^2}$, φ_π and φ_δ contribute to the conductance as

$$\bar{\sigma} = \sin^2\varphi_s + \sin^2\varphi_{3z^2-r^2} + 2\sin^2\varphi_\pi + 2\sin^2\varphi_\delta. \tag{33}$$

The electric conductance eq.(33) obviously takes a value between 0 and $6G_0$. The phase-shifts are, respectively, defined as

$$\cot\varphi_{s,3z^2-r^2} = \frac{-(a_1E_{3z^2-r^2} + a_2E_s) \pm \sqrt{(a_1E_{3z^2-r^2} - a_2E_s)^2 + 4a_3^2E_{3z^2-r^2}E_s}}{a_1a_2 - a_3^2}, \tag{34}$$

$$E_{3z^2-r^2} = \epsilon_{\mathrm{F}} - \epsilon_{3z^2-r^2}, \quad E_s = \epsilon_{\mathrm{F}} - \epsilon_s, \tag{35}$$

$$\cot\varphi_\pi = -\frac{\epsilon_{\mathrm{F}} - \epsilon_\pi}{b}, \tag{36}$$

$$\cot\varphi_\delta = -\frac{\epsilon_{\mathrm{F}} - \epsilon_\delta}{c}. \tag{37}$$

In this model, we impose the charge neutrality condition through the FSR as

$$10 = \frac{2}{\pi}(\varphi_s + \varphi_{3z^2-r^2} + 2\varphi_\pi + 2\varphi_\delta). \tag{38}$$

Considering eq.(32), the equations (34)-(37) take negative values. Therefore the phase-shifts have the following relationship.

$$\frac{\pi}{2} < \varphi_\delta < \varphi_\pi < \varphi_{3z^2-r^2} < \varphi_s < \pi \tag{39}$$

To elucidate the behavior of the electric conductance just before the Pt atomic contact fractures by the increment of the distance between electrodes, we discuss the Pt system under the condition that the distance between electrodes is larger than the equilibrium distance. When the distance increases, each phase-shift tends to π with the condition that they satisfy the relationship in eq.(39). Here we discuss the electric conductance in the range that three phase-shifts φ_s, $\varphi_{3z^2-r^2}$ and φ_π are regarded as nearly equal. Therefore the electric conductance and the FSR become

$$\bar{\sigma} = 4\sin^2\varphi_\pi + 2\sin^2\varphi_\delta, \tag{40}$$

$$10 = \frac{2}{\pi}(4\varphi_\pi + 2\varphi_\delta). \tag{41}$$

Finally, the electric conductance is represented as

$$\bar{\sigma} = 2\left(\sin\varphi_\delta - \frac{1}{2}\right)^2 + \frac{3}{2}. \tag{42}$$

Therefore the electric conductance of an atomic contact consisting of *sd*-orbital does not exceed $2G_0$ under the condition that the phase-shifts satisfy the FSR.

Furthermore we notice that when the matrix elements of Δ tend to 0 under the condition that the phase-shifts satisfy the FSR, there is the only allowable solution that φ_δ tends to $\pi/2$ and the other phase-shifts tend to π. This result is derived by a similar reason in the discussion about Al contact. Therefore the electric conductance increases towards $2G_0$. This result agree with the experiment involving Pt.[2]

4. Conclusion

We have analytically studied the electric conductance through nanostructure bridges consisting of an atom with *sp*-orbital and *sd*-orbital. We have analytically introduced the charge neutrality through the FSR. We have treated that increment of the distance between two electrodes is gradually separating interatomic distance. We suggest that the Friedel sum rule can leads to a guiding principle for interpretation of the phenomena in these contacts.

The electric conductance of an atomic contact consisting of atoms with *sp*-orbital such as Al can take a value from 0 to $4G_0$. However, we have shown that the electric conductance does not exceed $1G_0$ under the condition that the phase-shifts satisfy the FSR. Furthermore, when the matrix elements of Δ tend to 0 under the condition that the phase-shifts satisfy the FSR, the electric conductance increases to $1G_0$. In the same way, we have analytically shown the behavior of an atomic contact consisting of atoms with *sd*-orbital such as Pt. These results are consistent with the results of the experiments.[1,2]

We speculate that the expression given here can be applied even with the electron correlation effect, although we have not dealt with it. The effect would be involved in the actual calculation of the phase-shifts. This point must be studied in near future.

5. Acknowledgements

The authors acknowledge Y. Otsuka for valuable information about his previous works on the phase-shift analysis. Y.K. would like to thank T. Mii and K. Tsutsui for fruitful discussions. This work has been partially supported by the Grant-in-Aid for Scientific Research on Priority Areas No. 17069013 from the Ministry of Education, Culture, Sports, Science and Technology of Japan. The authors thank the Yukawa Institute of Theoretical Physics at Kyoto University, where this work was initiated [completed] during the YITP-W-10-12 on "International and Inter disciplinary Work on Novel Phenomena in Integrated Complex Science: from Non-Living to Living Systems".

Reference

[1] J. C. Cuevas, A. Levy Yeyati, A. Martin-Rodero, G. Rubio Bollinger, C. Untiedt, and N. Agrait, Phys. Rev. Lett. **81**, 2990 (1998).
[2] J. M. Krans, C. J. Muller, I. K. Yanson, Th. C. M. Govaert, R. Hesper, and J. M. van Ruitenbeek, Phys. Rev.B **48**, 14721 (1993).
[3] J. Friedel, Phil. Mag. **43** 153 (1952);
J. Friedel, Adv. Phys. **3** 446 (1954);
J. Friedel, Nuovo Cimento Suppl. **2** 287 (1958).
[4] S. Datta, W. Tian, Phys. Rev. B **55** R1914 (1997).

[5] K. Makoshi, N. Shima, Y. Otsuka, Physica E **29** 656 (2005).
[6] D.C. Langreth, Phys. Rev. **150** 516 (1966).
[7] P. Pernas, A. Martín-Rodero, and F. Flores, Phys. Rev. B **41**, 8553 (1990).
[8] J. C. Cuevas, A. Levy Yeyati, A. Martin-Rodero, G. Rubio Bollinger, C. Untiedt, and N. Agrait, Phys. Rev. Lett. **80**, 1066 (1998).
[9] K.T. Mahanthappa, Phys. Rev. **126** 329 (1961).
[10] P.M. Bakshi, K.T. Mahanthappa, J. Math. Phys. **4** 1 (1963).
[11] P.M. Bakshi, K.T. Mahanthappa, J. Math. Phys. **4** 12 (1963).
[12] L.V. Keldysh, Sov. Phys. JETP **20** 1018 (1965).
[13] Y.Meir, N.S.Wingreen, Phys.Rev. Lett. **68** 2512 (1992).
[14] K.Makoshi, T. Mii, Japan J.Appl.Phys. **34** 3325 (1995).

IIW-NPICS 2010 IOP Publishing
Journal of Physics: Conference Series **344** (2012) 012019 doi:10.1088/1742-6596/344/1/012019

Magnetic Properties of Itinerant Ferromagnet $La_{12}Fe_{57.5}As_{41}$

Bin Chen[1,2], JinHu Yang[1], Sho Takehara[1], Hiroto Ohta[1], Chishiro Michioka[1], Kazuyoshi Yoshimura[1]

[1] Department of Chemistry, Graduate School of Science, Kyoto University, Kitashirakawa Oiwake-cho, Sakyo-ku, Kyoto 606-8502, Japan
[2] Department of Physics, Hangzhou Normal University, Hangzhou 310036, China

E-mail: chenbin@kuchem.kyoto-u.ac.jp

E-mail: kyhv@kuchem.kyoto-u.ac.jp

Abstract. We successfully synthesized the single crystal of the itinerant ferromagnetic compound $La_{12}Fe_{57.5}As_{41}$ by Sn-flux method and studied its magnetic properties by magnetization measurements. The temperature dependence of magnetic susceptibility obeys the modified Curie-Weiss law, $\chi^{-1} = (\chi_0 + C/(T-\theta))^{-1}$, quite well in the high temperature region. The effective moment $P_{\rm eff}$ and the Weiss temperature θ are estimated to be 1.13 μ_B/Fe and 128.5 K, respectivelly. The magnetization of $La_{12}Fe_{57.5}As_{41}$ shows the convex behavior when plotted in a form of M^2 vs H/M at low temperatures, while the M^4 shows the linear relation against H/M in a wide temperature range near T_C.

1. Introduction

$La_{12}Fe_{57.5}As_{41}$ is a newly discovered itinerant ferromagnet with the Curie temperature $T_C = 125$ K[1]. It has the orthorhombic crystal structure with the space group of *Pmmn*. The complex three-dimensional network of $La_{12}Fe_{57.5}As_{41}$ is composed of novel wave-like layers of connected As-centered trigonal prisms[1]. The Fe atoms form the wave-like square lattice, which is parallel to the *ab* plane[1]. It is quite different with the FeAs layers in the mother compounds of Fe-based superconductors and the three-dimensional network in the $CaFe_4As_3$[2], both of which show the antiferromagnetic order at low temperatures. Therefore, it is very interesting to investigate the magnetic properties of $La_{12}Fe_{57.5}As_{41}$ due to the rare examples of ferromagnetic order in Fe-As contained compounds.

However, the ferromagnetic impurities are often introduced into the polycrystalline samples, which mask the intrinsic magnetic properties of $La_{12}Fe_{57.5}As_{41}$[1]. The high quality samples are strongly needed to elucidate the magnetic properties of this compound. In this work, we successfully synthesized the single crystal samples of $La_{12}Fe_{57.5}As_{41}$ and investigated its magnetic properties by the measurements of magnetic susceptibilities and isothermal magnetizations with the applied magnetic field parallel to the *ac* plane.

2. Experiments

The single crystal of $La_{12}Fe_{57.5}As_{41}$ was synthesized by the Sn-flux method. A 2 g mixture of La (Lump 99.9%), Fe (bulk 99.99%), and As powder (99.999%) in a molar ratio of 1:5:3 was

Published under licence by IOP Publishing Ltd

Journal of Physics: Conference Series **344** (2012) 012019 doi:10.1088/1742-6596/344/1/012019

placed in an evacuated silica tube together with 12 g of Sn (99.99%). The mixture was heated to 1050°C over 12 h, held at this temperature for 12 h, and then slowly cooled to 600°C with 3°C/h, followed by furnace cooling to room temperature. Black needle-shaped crystals of the ternary arsenides $La_{12}Fe_{57.5}As_{41}$, together with crystals of binary-phase byproduct FeAs, were gained after dissolving excess Sn flux in diluted acid HCl. The magnetization was measured using a superconducting quantum interference device magnetometer in the temperature range between 2 and 300 K with an applied magnetic field H up to 5 T.

3. Results and Discussion

Figure 1 shows the temperature dependence of the magnetization divided by the external magnetic field M/H (■) and the inverse magnetic susceptibility (□) for $La_{12}Fe_{57.5}As_{41}$ measured under $H = 0.5$ T with the magnetic field parallel to the ac plane. The ferromagnetic phase transition manifest itself by a rapid increase of the magnetic susceptibility at about 125 K with further decreasing the temperature, which is similar to the result of reference 1. However, the absolute magnitude of the susceptibility at low temperature is only about 25% of the value of susceptibility reported previously[1]. As we will point out below, this indicates that our sample is very pure without any magnetic impurities. On the other hand, the magnetic susceptibility obeys the modified Curie-Weiss law, $\chi^{-1} = (\chi_0 + C/(T - \theta))^{-1}$, quite well in high temperature region (150 K ~ 300 K) as shown in Fig. 1 by solid line. As a result, the temperature independent part χ_0, Curie constant C and Weiss temperature θ are estimated to be 0.016 emu/mol, 9.14 emu K/mol and 128.5 K, respectively. From the Curie constant C we obtained the effective moment $P_{\rm eff} = 1.13\ \mu_B$/Fe.

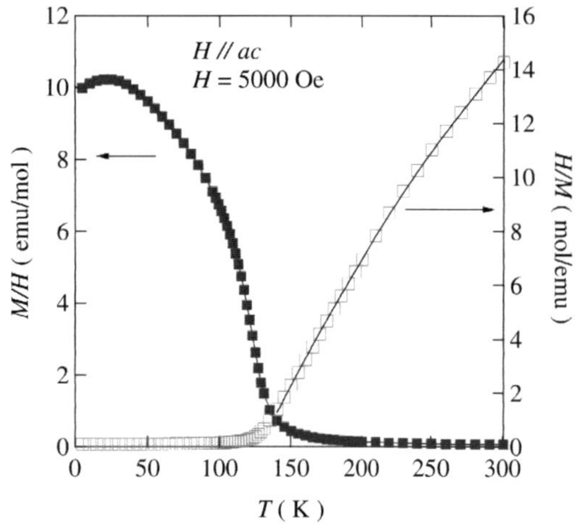

Figure 1. Temperature dependence of the magnetization divided by the external magnetic field M/H (■) and the inverse magnetic susceptibility (□) for $La_{12}Fe_{57.5}As_{41}$ with the external magnetic field applied parallel to the ac plane.

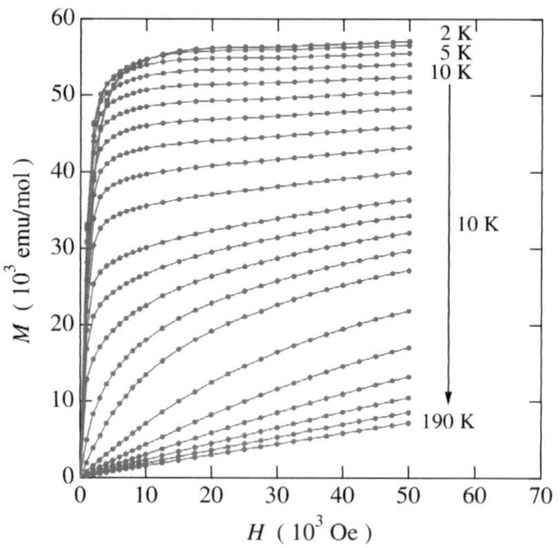

Figure 2. $M(H)$ at various temperatures with the magnetic field applied parallel to the ac plane.

Figure 2 shows the isothermal magnetization curves of $La_{12}Fe_{57.5}As_{41}$ from 2 K to 190 K under the external magnetic field up to 5 T. The magnetization saturates in the field of 2 T at 2 K. Moreover, with the increase of temperature, the saturated magnetization decreases continuously in the ferromagnetic ordered state. The linear relationship between M and H

can be found at the temperature higher than 170 K. Our results are quite different with that reported by S. S. Stoyko *et al.*, where the linear relation is absent even in the paramagnetic state with the temperature up to 300 K[1]. This is because of the existence of the magnetic impurities in their sample as they claimed. These results suggest our sample is of high quality.

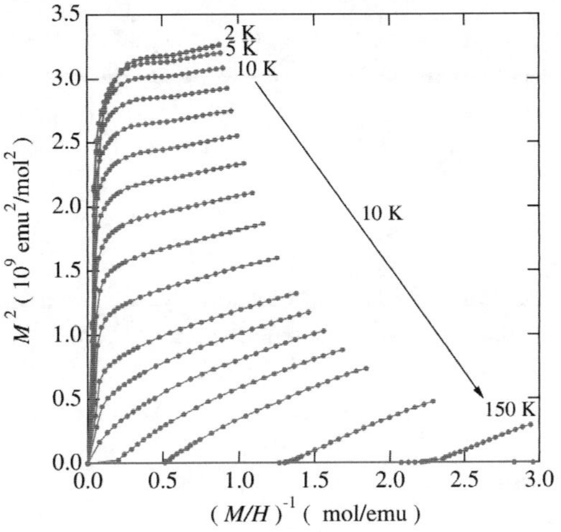

Figure 3. Arrott Plot (M^2 versus H/M plot) for $La_{12}Fe_{57.5}As_{41}$ at various temperatures.

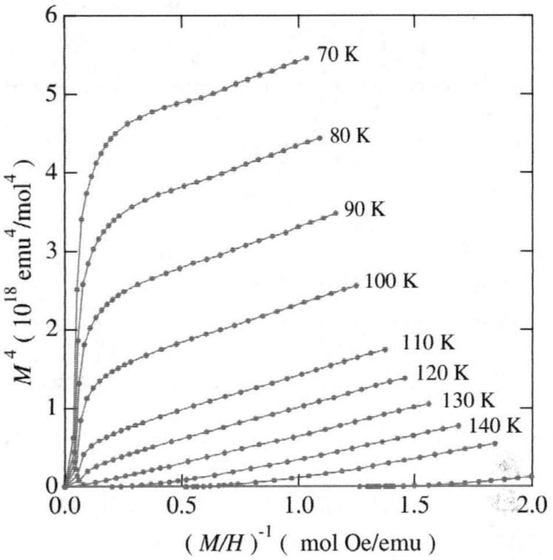

Figure 4. M^4 versus H/M plot for $La_{12}Fe_{57.5}As_{41}$ at various temperatures.

On the basis of the Stoner model, the high field isothermal magnetization follows $M^2 = A + B(H/M)$, where the parameters A and B are independent of magnetic field H. This leads to a series of linear and parallel curves in M^2 versus H/M plots that are commonly so-called Arrott plot. This behavior has been observed in some magnetic materials, such as $ZrZn_2$[3], Y(Co-Al)$_2$[4, 5], Ni_3Al[6], Sc_3In[7]. Figure 3 shows the M^2 vs H/M plots at various temperatures from 2 K to 150 K. The linear relationship between M^2 and H/M can be found with the temperature lower than 80 K. Usually, the Curie temperature T_C can be determined at the temperature, where the intercept along the M^2 axis in the Arrott plot crosses the origin of the coordinate axes, whereas, in the temperature range between 90 K and 130 K, the M^2 does not show the linear relation but a convex curvature against H/M. It is difficult to determine the T_C from the Arrott plot in this case. However, if we plot them in a M^4 vs H/M form, it is clear that M^4 shows good linear relation with H/M in a wide temperature range near T_C in $La_{12}Fe_{57.5}As_{41}$ as shown in Fig. 4. This behavior has been observed in some weak itinerant ferromagnetic materials, such as MnSi[8], LaCoAsO[9]. The linear relationship of M^4 with H/M is explained on Takahashi's theory of spin fluctuations, which assumed that the sum of zero-point and thermal spin fluctuations is independent of temperature[10]. Therefore, T_C is estimated to be 125 K approximately from the M^4 vs H/M plot.

In order to clarify a magnetic material belongs to the itinerant or a localized spin system, we estimate the Rhodes-Wohlfarth ratio of $La_{12}Fe_{57.5}As_{41}$, which is defined as the value of P_c/P_s, where $P_c(P_c+2) \equiv P_{\rm eff}^2$ and P_s is the spontaneous magnetization at ground state[11]. In the case of the weak itinerant ferromagnet, Rhodes-Wohlfarth ratio is known to be much larger than the case of the localized spin system $P_c/P_s \equiv 1$. We estimate $P_c = 0.51\mu_B$/Fe by using the value $P_{\rm eff} = 1.13\ \mu_B$/Fe. $P_s = 0.17\ \mu_B$/Fe is estimated from the Arrott plot as shown in Fig. 3. Using these values we obtained P_c/P_s of $La_{12}Fe_{57.5}As_{41}$ is about 3, larger than the 1, which indicates this material belongs to the itinerant ferromagnet class.

4. Summary

The itinerant ferromagnetic compound $La_{12}Fe_{57.5}As_{41}$ was synthesized successfully by Sn-flux method. The magnetic properties were studied by the measurements of magnetic susceptibility and magnetization with the external magnetic field applied in the *ac* plane. The temperature dependence of magnetic susceptibility obeys the modified Curie-Weiss law, $\chi^{-1} = (\chi_0 + C/(T - \theta))^{-1}$ quite well in high temperature region. The effective moment P_{eff} and Weiss temperature θ are estimated to be 1.13 μ_{B}/Fe and 128.5 K, respectively. We showed that the magnetization of $La_{12}Fe_{57.5}As_{41}$ has the linear behavior when plotted in the form of M^2 vs H/M in the lower and the higher temperature regions, while changes to the M^4 vs H/M in a wide temperature range near T_{C}. The Rhodes-Wohlfarth ratio is estimated to be about 3, suggesting that $La_{12}Fe_{57.5}As_{41}$ is an itinerant ferromagnet.

5. Acknowledgments

The authors thank the Yukawa Institute for Theoretical Physics at Kyoto University. Discussions during the YITP workshop YITP-W-10-12 on "International and Interdisciplinary Workshop on Novel Phenomena in Integrated Complex Sciences: from Non-living to Living Systems" were useful to complete this work. This workshop was supported in part by the Grant-in-Aid for the Global COE Program "The Next Generation of Physics, Spun from Universality and Emergence" from the Ministry of Education, Culture, Sports, Science and Technology (MEXT) of Japan. This work is also supported by a Grant-in-Aid for Scientific Research on Priority Areas, "Invention of anomalous quantum materials", from the Ministry of Education, Culture, Sports, Science and Technology of Japan (16076210), and by Grants-in-Aid for Scientific Research from the Japan Society for the Promotion of Science (19350030, 22350029).

References

[1] Stanislav S Stoyko, Peter E R Blanchard and Arthur Mar 2010 *Inorg. Chem.* 49 2325-33
[2] I Todorov, D Y Chung, C D Malliakas, Q A Li, T Bakas, A Douvalis, G Trimarchi, K Gray, J F Mitchell, A J Freeman and M G Kanatzidis 2009 *J. Am. Chem. Soc.* 131 5405-07
[3] E A Yelland, S J C Yates, O Taylor, A Griffiths, S M Hayden and A Carrington 2005 *Phys. Rev. B* 72 184436
[4] K Yoshimura, M Takigawa, Y Takahashi, H Yasuoka and Y Nakamura 1987 *J. Phys. Soc. Jpn.* 56 1138
[5] K Yoshimura and Y Nakamura 1985 *Solid State Communications.* 56 767
[6] B Chen, C Michioka, Y Itoh and K Yoshimura 2008 *J. Phys. Soc. Jpn.* 77 103708
[7] T Hioki and Y Masuda 1977 *J. Phys. Soc. Jpn.* 43 1200
[8] M K Chattopadhyay, Parul Arora and S B Roy 2009 *J. Phys.: Condens. Matter* 21 296003
[9] H Ohta, K Yoshimura 2009 *Phys. Rev. B* 79 184407
[10] Y Takahashi 1986 *J. Phys. Soc. Jpn.* 55 3553
[11] P R Rhodes and E P Wohlfarth 1963 *Proc. Roy. Soc. London* 273 247

Probing Onset of Coherence in $CeIrIn_5$

A C Shockley, A P Dioguardi, N apRoberts-Warren, P Klavins, N J Curro
Physics Department, University of California Davis,
One Shields Ave Davis, CA, USA, 95616

E-mail: shockley@student.physics.ucdavis.edu

Abstract. We report ^{115}In NMR measurements in the heavy fermion superconductor $CeIrIn_5$. The Knight shift for the symmetric In(1) site displays the same behaviour as the related material $CeCoIn_5$. Our results support the phenomenological two-fluid model proposed by Nakatsuji et al. Comparison to recent DMFT calculations for $CeIrIn_5$ further strengthens our conclusions.

1. Introduction

Since the discovery of the heavy fermion class of superconductors, scientists have been trying to understand their microscopic electronic behaviour. Among families such as the $CeMIn_5$ (M = Ir, Co, Rh) class, research focuses around what drives one compound, e.g. $CeRhIn_5$, to be anti-ferromagnetic [1] and another compound, e.g. $CeIrIn_5$ or $CeCoIn_5$, to be superconducting [2,3,4] at ambient pressure. The question remains open as these materials are notoriously difficult to model theoretically. Throughout this class, however, certain normal state properties appear to be universal. Nakatsuji et al [5] (henceforth NPF) proposed that this scaling behaviour could best be explained by a two-fluid model. At a temperature T*, these compounds possess itinerant behaviour; the heavy-electron itinerant component, described by an order parameter f(T/T*), scales with T*. This phenomenon is best illustrated by the Knight Shift anomaly. In this paper, we present data for the Knight Shift anomaly in $CeIrIn_5$ and compare this data with the behaviour of $CeCoIn_5$.

2. Knight Shift Anomaly

The Kondo lattice can be modelled as a lattice of local f-moments embedded in a sea of conduction electrons. In the two-fluid model proposed by NPF, a fraction of the f-moments, f(T), delocalize below a temperature T* to form a coherent state with the conduction electrons much like in superfluid ^{4}He. The magnetic system then has one component corresponding to the localized f-moments with susceptibility χ_{KI} and one component for the itinerant heavy-electrons with susceptibility χ_{HF},

$$\chi(T) = [1 - f(T)]\chi_{KI}(T) + f(T)\chi_{HF}(T) \tag{1}$$

The total spin of the system is a combination of the localized f-moments and the conduction electron spins:

$$S_{tot} = \sum_i \boldsymbol{S}^f(\boldsymbol{r}_i) + \sum_l \boldsymbol{S}^c(\boldsymbol{r}_l), \tag{2}$$

where r_i is the location of the local f-moments and r_l is the location of the itinerant component [6]. The susceptibility is represented by the correlation function of the spins so that the uniform susceptibility is:

$$\chi = \chi_{ff} + 2\chi_{cf} + \chi_{cc} \approx \chi_{ff} + 2\chi_{cf} \tag{3}$$

where χ_{ff} = 1/N $<\mathbf{S}^f(\mathbf{r}_i)\mathbf{S}^f(\mathbf{r}_{i'})>$ and χ_{cf} = 1/N $<\mathbf{S}^c(\mathbf{r}_l)\mathbf{S}^f(\mathbf{r}_i)>$ are the orbital resolved susceptibilities. Since the susceptibility from the background conduction electrons is small, we omit it to obtain the NPF result. In principle, these two components may have different temperature dependences.

In general, the hyperfine couplings to these two spin components will differ [6]. We write the hyperfine Hamiltonian as:

$$\mathcal{H}_{hyp} = \gamma\hbar \textstyle\sum_l \boldsymbol{I}(\boldsymbol{r}_l) \cdot \boldsymbol{A} \cdot \boldsymbol{S}^c(\boldsymbol{r}_l) + \gamma\hbar \textstyle\sum_i \boldsymbol{I}(\boldsymbol{r}_i) \cdot \boldsymbol{B} \cdot \boldsymbol{S}^f(\boldsymbol{r}_i) \tag{4}$$

where **A** and **B** are temperature independent contact and transferred hyperfine couplings, respectively, and $\mathbf{r}_i$ are the locations of the nearest neighbour 4f sites. The Knight shift is related to the hyperfine couplings by

$$K_\alpha(T) = K_{0,\alpha} + (A_\alpha + B_\alpha)\chi_{cf}(T) + B_\alpha \chi_{ff}(T) \tag{5}$$

where $K_{0,\alpha}$ is an offset and we have dropped the summation over the neighbouring sites for simplicity. For T>T*, we assume that $\chi_{cf}(T) \sim 0$ so that we can solve for the coupling constant B. Below T*, $K_\alpha(T)$ and $\chi(T)$ are no longer proportional. $\chi_{cf}(T)$ and $\chi_{ff}(T)$ enter into $\chi(T)$ and $K_\alpha(T)$ with different weights due to the onset of coherence.

In order to investigate the onset of coherence, we have performed NMR measurements on a single crystal of $CeIrIn_5$. A 2x2x1 mm^3 single crystal was grown according to the standard growth

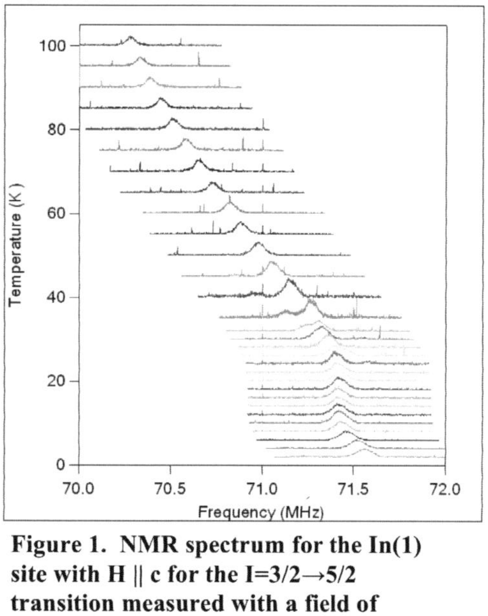

Figure 1. NMR spectrum for the In(1) site with H || c for the I=3/2→5/2 transition measured with a field of 79.351kG. Spectra are shifted to correspond to the temperature at which they were measured.

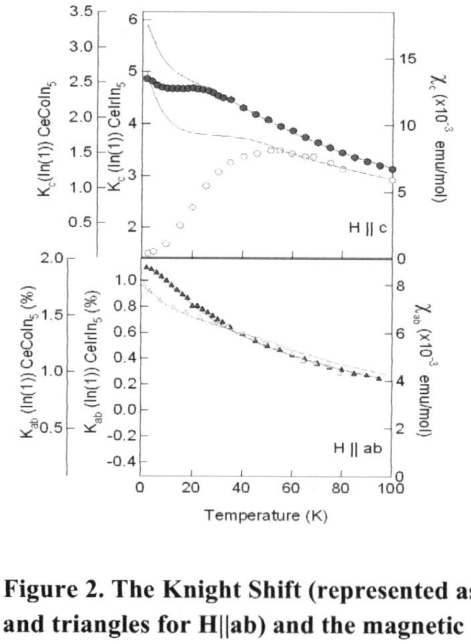

Figure 2. The Knight Shift (represented as circles for H||c and triangles for H||ab) and the magnetic susceptibility (represented as solid lines) are plotted as a function of temperature. The data for $CeIrIn_5$ are shown as solid blue for H||c and purple for H||ab. The data for $CeCoIn_5$ are shown as empty pink for H||c and green for H||ab [8].

method [7] and acid-etched to remove excess metallic indium from the surface. $CeIrIn_5$ has two unique crystallographic ^{115}In sites (I=9/2); we measured the axially symmetric In(1) site. In a Quantum Design PPMS, we acquired field-swept spectra with the crystalline c-axis oriented both perpendicular and parallel to the applied field. As a function of temperature, we measured the shift of the I=3/2□5/2 transition with an applied field of 79.351kG parallel to the c-axis shown in Fig 1 and of 66.209kG perpendicular to the c-axis.

The onset of coherence, T*, is manifest as a breakdown in the linear relationship between K and χ (see Figs 2, 3). For the field applied parallel to the c-axis, we found a hyperfine coupling constant of 25.6 ± 0.2 kG/μ_B from the linear fit shown in Fig 3. The contribution K_{cf} can be found by subtracting our data from the linear fit shown in Fig 4. To obtain a value for T*, we compare our data to the NPF model [9]:

$$K_{cf} = K_{cf0}(1 - \frac{T}{T^*})^{\frac{3}{2}}(1 + \ln\frac{T}{T^*}) \qquad (6)$$

where K_{cf0} is a scaling factor. According to this model, T* for $CeIrIn_5$ is 34.9 ± 1 K.

3. Discussions and Conclusion

Recent work by Kambe et al on the Knight shift in $CeIrIn_5$ matches the trend of the data presented in this paper but yields a lower hyperfine coupling constant, B [10]. They argue that the two-fluid model gives unphysical results for the magnetic fluctuations measured by the anisotropy of the nuclear spin lattice relaxation rate (SLRR). On the other hand, a two component analysis of the SLRR in $CeCoIn_5$ yields new insight into the origin of superconductivity in this system [11]. Based on the similar behaviours of both these materials as presented here, we would expect that the NPF model would either physically represent both $CeIrIn_5$ and $CeCoIn_5$ or neither. Currently, this question is still open to debate. In comparison with DMFT + LDA calculations for $CeIrIn_5$, a slow build-up of coherence is predicted as in the NPF approach [12]. In conclusion, the key to a microscopic understanding of the $CeMIn_5$ materials may be found by exploiting their similarities rather than highlighting their uniqueness.

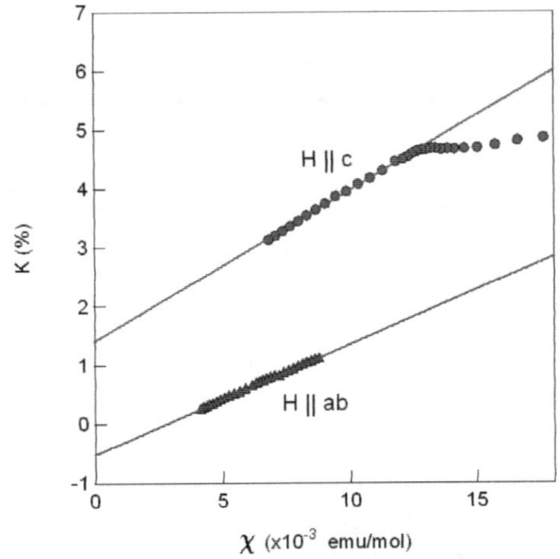

Figure 3. The Knight Shift versus the susceptibility for the In(1) site of $CeIrIn_5$ with temperature as an implicit parameter. Symbols are the same as in Fig 2. The high temperature data was fit to a linear regression to obtain the hyperfine coupling constant, B.

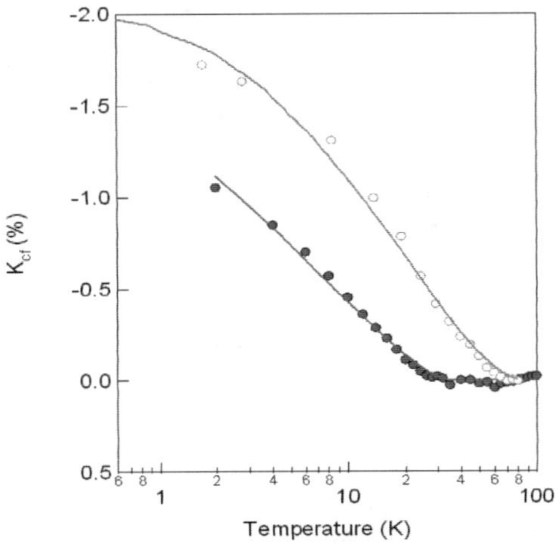

Figure 4. K_{cf}, the difference between the data and the linear regression from Fig 3, is plotted as a function of the log of the temperature. Symbols are the same as in Fig 2. Solid lines are fits to Eq 6 for $CeIrIn_5$, blue, and $CeCoIn_5$, pink.

Acknowledgments
The authors thank the Yukawa Institute for Theoretical Physics at Kyoto University. Discussions during the YITP workshop YITP-W-10-12 on "International and Interdisciplinary Workshop on Novel Phenomena in Integrated Complex Sciences: from Non-living to Living Systems" were useful to complete this work. The authors would also like to thank the conference organizers as well as ICAM/I2CAM for travel funding. We also thank Yi-Feng Yang and David Pines for enlightening discussions.

References
[1] H. Hegger, et al., Phys. Rev. Lett. **84** (2000) 4986.
[2] C. Petrovic et al., Europhys. Lett. **53** (2001) 354
[3] C. Petrovic et al., J. Phys. Condens. Matter **13** (2001) L337
[4] R Movshovic et al., Physica B **312-313**, (2002) 7-12.
[5] S Nakatsuji et al., Phys. Rev. Lett. **92**, 016401 (2004).
[6] NJ Curro et al., Phys. Rev. B **70**, 235117 (2004).
[7] Moshopoulou et al., Journal of Solid State Chemistry, Vol 158, Issue 1, 2001, P 25-33.
[8] NJ Curro et al., Phys. Rev. B **64**, 180514(R) (2001).
[9] Yi-Feng Yang and David Pines, PRL **100**, 096404 (2008).
[10] S. Kambe et al., Phys. Rev. B **81**, 140405(R) (2010).
[11] Yi-Feng Yang et al., PRL **103**, 197004 (2009).
[12] JH Shim, K Haule and G Kotliar, *Science* **318**, 1615 (2007).

Journal of Physics: Conference Series **344** (2012) 012021 doi:10.1088/1742-6596/344/1/012021

Understanding of Superconductivity in Elemental Substances by Pseudopotential Radii

Y. Makino and K. Yoshimura

Graduate School of Science, Kyoto University, Kitashirakawa Oiwake-cho, Sakyo-ku, Kyoto, 606-8502, Japan

E-mail: ymak@kuchem.kyoto-u.ac.jp

Abstract. Two dimensional map on superconductivity in elemental substances was constructed using pseudopotential radii and its derived parameter, $(Z/r(s))^{1/2}$, where $r(s)$ and Z are the pseudopotential radius of s electron and valence. It is found that superconducting elements are well mapped by selecting the difference between pseudopotential radii and $(Z/r(s))^{1/2}$ as the coordinates. It is suggested that transition temperature(T_c) of superconductivity is related to the $(Z/r(s))^{1/2}$ and the cube of r(s) except several transition metals.

1. Introduction

Since the discovery of high T_c cuprate superconductors[1], it has been continued to search the controlling factors of high T_c and other new superconductors[2-4]. It has been indicated that electronegativity is a key chemical parameter for understanding superconductivity in both elemental substances and compounds[5,6]. For example, low-T_c superconductors have a value of electronegativity ranging from 1.2 to 2.32 in Pauling's scale and high-T_c superconductors are plotted around a value of electronegativity(for example, 2.5 in Pauling's scale)[7]. Similar results have been reported in the other approaches based on the electronegativity[8,9]. In these approaches, however, the relation between physical picture and electronegativity, which is a typical chemical parameter, has not been clarified well.

Pseudopotential radius is a physical parameter corresponding to the radius which pseudopotential for respective valence electron (that is, s, p or d electron) becomes zero in atomic system with many electrons[10] and has been applied to construct the crystal structural maps for compounds[11,12] and to phase designing of hard coatings[13,14]. Pseudopotential radius of s electron has been connected with the electronegativity and the orbital electronegativity for s electron can be defined by the pseudopotential radius for respective s electron through the formulae of $\chi(s)=k(Z/r(s))^{1/2}$[15], where $\chi(s)$:orbital electronegativity of s electron, k: constant, Z:valence, $r(s)$:pseudopotential radius of s electron. Thus, pseudopotential radius is the interesting parameter that can combine physical picture with chemical characters. In this study, it is aimed to construct a two dimensional map on the superconducting elemental substances by using the pseudopotential radius and orbital electronengativity in order to understand superconductivity from chemical standpoint. Further, it is tried to describe transition temperature(T_c) in elemental substances by pseudopotential radius.

2. Selection of parameters

Pseudopotential radii calculated by Zunger are used in this study because the radii of s, p and d electrons for 70 elements are available[16]. Under the assumption that the electrons in the outermost Fermi surface contribute to superconductivity, the root of the valence divided by pseudopotential radius of s electron (*r(s)*, that is $(Z/r(s))^{1/2}$, is adopted as the abscissa in the two dimensional map. Differences between the pseudopotential radii of s electron and p electron or between s electron and d electron, that is $|r(s)-r(p)|$ or $|r(s)-r(d)|$, are taken as the ordinate. In this study, orbital electronegativity of s electron for respective element is approximately adopted by the value of $(Z/r(s))^{1/2}$. In order to examine the dependence of T_c on the pseudopotential radius of s electron, it is assumed that s and p electrons contribute to superconductivity in so-called sp element and two s electrons simply contributes to superconductivity in transition metal. All data on T_c are quoted from Ref.17.

3. Results and Discussion

Two dimensional map by taking $(Z/r(s))^{1/2}$ and $|r(s)-r(p)|$ (or $|r(s)-r(d)|$) as coordinates is shown in Fig.1. In this map, the data are plotted by distinguishing among superconductor(SC), metastable superconductor(meta-) and non-superconductor(non-SC). The values of $|r(s)-r(d)|$ are adopted as the ordinate values for three post-transition metals(that is, Cu, Ag and Au). As shown in this figure, decisive factor for superconductivity is the orbital electronegativity of s electron and the sp-bonded elemental superconductors are plotted between $(Z/r(s))^{1/2}\approx 1.3$(lower limit) and $(Z/r(s))^{1/2}\approx 2.3$. In the transition metals, superconductivity appears beyond the threshold value of $(Z/r(s))^{1/2}$ in sp-bonded elemental substances and this may be due to d-band in the transition metals. The threshold value for superconductivity of transition metals in the larger $(Z/r(s))^{1/2}$ side decreases with increasing $(Z/r(s))^{1/2}$.

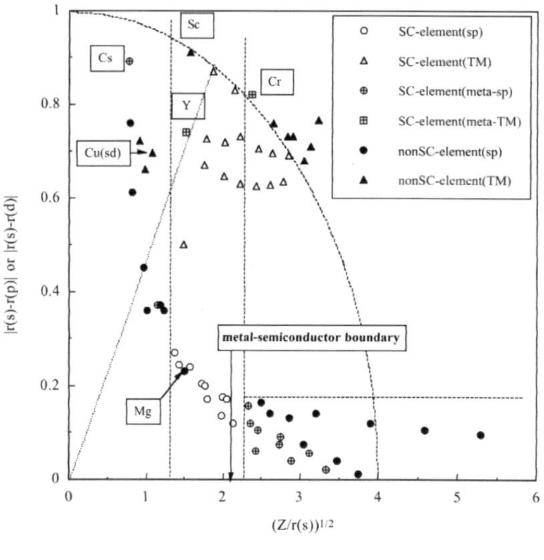

Figure 1. Two dimensional map on the superconductivity for elemental substances by taking $(Z/r(s))^{1/2}$ and $|r(s)-r(p)|$ (or $|r(s)-r(d)|$) as the ordinates. (*r(s)*, *r(p)* and *r(d)* are the pseudopotential radii, respectively, and Z is valence.)

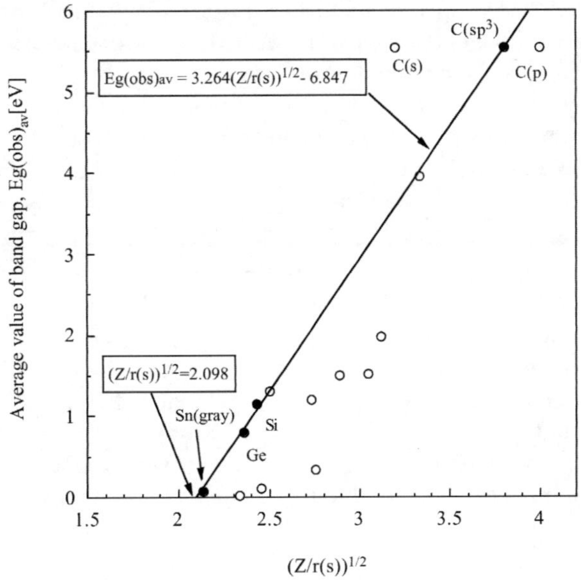

Figure 2. Dependence of average band gap in elemental substances on the $(Z/r(s))^{1/2}$.

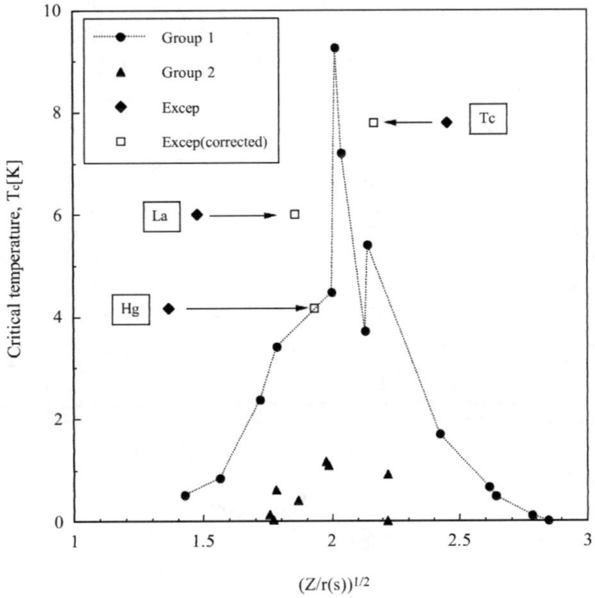

Figure 3. Relation between transition temperature(T_c) and $(Z/r(s))^{1/2}$. (The values of $(Z/r(p))^{1/2}$ is used as corrected plots for La and Tc. The value of $\sqrt{2}\,(Z/r(s))^{1/2}$ is adopted as the corrected plot for Hg.

Subsequently, we obtain Fig.2 by examining the relation between band gap of some elemental substance and $(Z/r(s))^{1/2}$. From fig.2, we can easily understand that the threshold of metal-semiconductor boundary is given by the $(Z/r(s))^{1/2}=2.098$ and thus superconductivity generates in the sp-bonded elemental substances around the threshold value irrespective of whether metal or semiconductor. The threshold for sp-bonded elements may be connected with the phenomenon that the overlapping of s-band and p-band in metallic elements is resolved and new gap is just about to form between these bands, though not simply explained in the complex band structures. The lower limit of superconductivity in $(Z/r(s))^{1/2}$ scale may be related to the electron density though details remain unclear. Several transition metals with a value of $(Z/r(s))^{1/2}$ larger than the threshold for sp-bonded elements show superconductivity. Detailed reason is unclear but similar electronic state in Fermi surface to that in semi-conducting sp-bonded elements seems to be formed by the existence of d-band. Higher limit for transition metals depends on the both $(Z/r(s))^{1/2}$ and difference between $r(s)$ and $r(d)$ electrons, that is $|r(s)-r(d)|$, and the higher limit is tentatively shown by a curve in Fig.1. At present, there is no method except to empirically determine the functional form of the higher limit by experimental data on the compounds with the contribution of d electron. Difference between $r(s)$ and $r(p)$, that is $|r(s)-r(p)|(=\Delta r(sp))$, are often used as an important coordinate in the structural mapping[11,12]. The physical meaning of $\Delta r(sp)$ has been not clearly explained but it has been suggested that $\Delta r(sp)$ is related to sp hybridization or orbital nonlocality as indicated in the structural map of AB compounds[10,18]. Likewise, the physical meaning of $|r(s)-r(d)|$ may be connected with sd hybridization or orbital nonlocality between s and d electrons.

When the straight line with $(Z/r(s))^{1/2}\approx 1.3$ is the lower side boundary for superconductors, Mg and Sc are plotted in the domain of superconductive elements though superconductivity have not been reported in these elements. In these elements and compounds including these elements, however, superconductivity can be probably generated because some compounds such as MgB_2 show superconductivity. Further, contamination, especially of oxygen, should be considered in the active elemental substances such as Sc and Y.

Examining the dependence of T_c against $(Z/r(s))^{1/2}$, Fig.3 is obtained. Excluding three elemental substances(La, Hg and Tc), the elemental substances are divided into two groups. In the one group (Group 1), the T_c value depends on $(Z/r(s))^{1/2}$ and shows the maximum for the elemental substances (such as Nb and Pb) with a $(Z/r(s))^{1/2}$ value ($\approx$2.0) close to the threshold value(=2.098) for metal-semiconductor boundary. In other group (Group 2), clear dependence is not shown against $(Z/r(s))^{1/2}$ value and many early transition metals belong to this group. The plots of La and Tc are fairly well placed close to the curve for Group 1 if p-character is predominant. The plot of Hg is well placed by assuming the formation of dimer.

An interesting result is obtained as shown in Fig.4 when the dependence of $T_c/N(e)$ ($N(e)$; number of valence electron) against $r(s)^3$ is examined. Superconducting elemental substances are distinguished into three groups and two groups are linearly plotted, respectively, except for several transition metals belonged to third group. The value of $N(e)$ is easily determined in the sp-bonded elements but the values are not determined in the transition metals. Here, data for transition metals are plotted in Fig.4 by giving $N(e)$=2 tentatively. It is found in Fig.4 that three transition metals (Nb, Tc and V) are well placed in the linear relation obtained from sp-bonded elements. The findings suggest that electronic character related to superconductivity of these transition metals is similar to that of sp-bonded elements. That is, superconductivity for these elements is generated by outermost sp electrons and is approximately controlled by s electrons. Because the interatomic distances of sp-bonded elements are linearly correlated with $r(s)$ (more precisely with the weighted average of $r(s)$ and $r(p)$)[19], it is interpreted that the physical meaning of $r(s)^3$ is approximately the spatial volume occupied by s electrons. On the other hand, La, Ta, Tl and Cd are placed in other straight line but the plots of these elements can be placed close to the linear relation obtained from sp-bonded elements when the $N(e)$ value is taken as $N(e)$=1. According to the result by Nieminen and Hodge[20], it has been reported that the sum of s and p electron (or sp electrons) numbers is between 1.0 and 2.0 though the sum has not been calculated for all transition metals. In this study, tentative adoption of $N(e)$=2 as the number of valence electron for transition metals may be reasonable as far as s electron is mainly the outermost one in Fermi surface though the tentative adoption is arguable. There is no clear evidence on reduction of $N(e)$ value to 1 for La, Ta, Tl and Cd. High stability of $(6s)^2$ by relativistic effect, especially in 5d transition metals, may be connected with reduction of the effective number of valence electron as the stable formation of monovalent state is indicated in Tl[21]. Conclusively, it is suggested that the effective number of electrons related to superconductivity is probably different in between Group 1 and Group 2. The elements belonged to Group 3 show no dependence against $r(s)^3$ probably by two reasons; (1) the strong contribution of d electron, (2) small number of electrons related to superconductivity. Reason (1) is closely connected with bond character and the electronic state can be changed with changing cohesion state, for example from bulk to film. Taking Mo and W as the examples, the plots of Mo(film) and W(film) are placed in the straight lines of Group 1 and Group 2, respectively, because the T_cs of Mo(film) and W(film) are 8K and 4.1K, respectively. Thus, the electronic state related to superconductivity can be changed by cohesion state. Further, the influence of impurity has been suggested in the superconductivity of film[22].

In near future, we will reported the two dimensional map on both superconductive elements and some compounds by similar coordinates based on $(Z/r(s))^{1/2}$. Further, it should be noted that $(Z/r(s))^{1/2}$ and $r(s)$ are not best parameter for understanding of superconductivity because superconductivity is not always controlled by the electronic state of s electron alone. In further investigation, detailed understanding of superconductivity in both elemental substances and compounds will be reported by constructing modified parameters including pseudopotential radius of p electron.

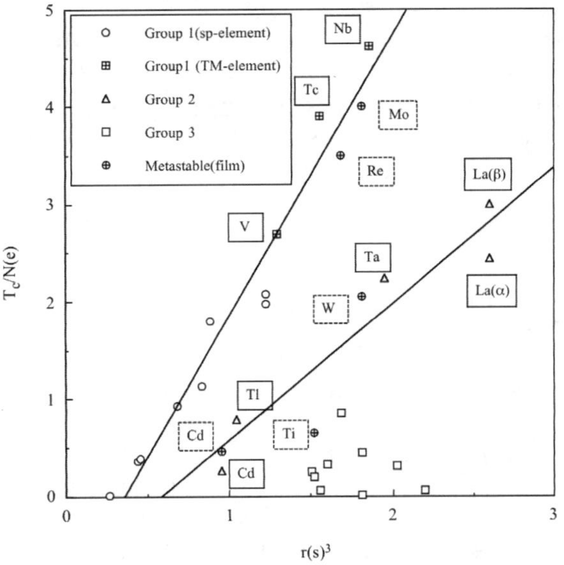

Figure 4. Dependence of $T_c/N(e)$ on the cube of $r(s)$. ($N(e)$ is the number of valence electrons and it is assumed to be tentatively $N(e)$=2 for all transition metals.

4. Summary

It was tried to construct a two dimensional map based on the pseudopotential radii and its derived parameter, $(Z/r(s))^{1/2}$, where $r(s)$ and Z are the pseudopotential radius of s electron and valence, in order to understand the superconductivity of elemental substances. It is found that superconducting elements are well mapped by selecting the difference between pseudopotential radii and $(Z/r(s))^{1/2}$ as the coordinates. Superconductivity in sp-bonded elements appears between $(Z/r(s))^{1/2}\approx1.3$ and $(Z/r(s))^{1/2}\approx2.3$. Several transition metals with a value of $(Z/r(s))^{1/2}$ larger than the higher threshold for sp-bonded elements show superconductivity. It is indicated that the metal-semiconductor boundary for sp-bonded elements appears around $(Z/r(s))^{1/2}=2.0$. With approaching to the boundary, further, superconducting element tends to show a higher T_c value even in the transition metals except several elements. Two linear relations between T_c and the cube of $r(s)$ are obtained, except several transition metals, under the assumption that superconductivity attributes to s and p electrons in sp-bonded elements and to two s electrons in transition metals, though the contribution due to d electrons should be considered in the exceptional transition metals.

Acknowledgements:

The authors thank the Yukawa Institute for Theoretical Physics at Kyoto University. Discussions during the YITP workshop YITP-W-10-12 on "International and Interdisciplinary Workshop on Novel Phenomena in Integrated Complex Sciences: from Non-living to Living Systems" were useful to complete this work. This workshop was supported in part by the Grant-in-Aid for the Global COE Program "The Next Generation of Physics, Spun from Universality and Emergence" from the Ministry of Education, Culture, Sports, Science and Technology (MEXT) of Japan.

References:

[1] Bedonorz J G and Müller K A 1986 *Z. Phys. B* **64** 189-193
[2] Mueller K A 2006 *Supercond. Sci. Technol.*, **19** S1-S3
[3] Geballe T H 2006 *J. Supercond. Novel Magnetism* **19** 261-276
[4] Phillips J C 2009 *Chem. Phys. Letts.* **473** 274-278
[5] Luo Q -G and Wang R -Y 1987 *J. Phys. Chem. Solids* **48** 425-430
[6] Jou C J and Washburn J 1991 *Appl. Phys.* **A53** 87-93
[7] Buzea C and Yamanshita T 2000 *Physica B* **281&282** 951-952
[8] Asokamani R and Manjula R 1989 *Phys. Rev. B* **39** 4217-4221
[9] Ichikawa S 1989 *J. Phys. Chem. Solids* **50** 931-934
[10] Zunger A 1981 in *Structue and Bonding in Crystals* **1** *ed. by M.O'Keeffe and A.Navrotsky* (New York: Academic Press)
[11] Villars P 1983 *J. Less-Common Metals* **92** 215-238
[12] Villars P, Mathis K and Hullinger F "Cohesion and Structure", Vol.2 (Ed. by F.R.Boer and F.R.Pettifor), Chap.1, p.1-103
[13] Makino Y 1998 *ISIJ International* **38** 925-934
[14] Makino Y 2005 *Surf. Coat. Technol.* **193** 185-191
[15] Zhang S B and Cohen M L 1989 *Phys. Rev.* **39** 1077-1080
[16] Zunger A 1980 *Phys. Rev. B* **22** 5839-5872
[17] Lide D R 2008-2009 *CRC Handbook of Chemistry and Physics* 12-58
[18] John J St and Bloch A N 1974 *Phys. Rev. Lett.* **33** 1095-1098
[19] Makino Y 1995 *J. Alloys and Compounds* **227** 18-27
[20] Nieminen R M and Hodges C H 1976 *J. Phys.: Metal Phys.* **6** 573-585
[21] Pyykko P 1988 *Chem. Rev.* **88** 563-594
[22] Johnson W L 1979 *J. Appl. Phys.* **50** 1557-1563

Local magnetic inhomogeneities observed via ^{75}As NMR in $Ba(Fe_{1-x}Ni_x)_2As_2$ with $H_0 \perp$ c-axis

A P Dioguardi[1], **N apRoberts-Warren**[1], **A C Shockley**[1], **S L Bud'ko**[2], **N Ni**[2], **P C Canfield**[2], **and N J Curro**[1]

[1]Department of Physics, University of California Davis
One Shields Ave., Davis, CA 95616

[2]Ames Laboratory & Iowa State University
Ames, IA 50011

E-mail: dioguardi@ms.physics.ucdavis.edu

Abstract. We present field-swept ^{75}As Nuclear Magnetic Resonance (NMR) measurements in the spin-density wave (SDW) state of $Ba(Fe_{1-x}Ni_x)_2As_2$ for x=0.0072. The large single crystals were aligned with the external field H_0 perpendicular to the c axis. The spectra are increasingly broadened as a function of doping, and are well fit by a model of a commensurate SDW with local impurities.

1. Introduction

The discovery of high temperature superconductivity by Bednorz and Müller in 1986[1] raised both hopes and questions about the nature of the Cooper pairing mechanism. The inability of phonon pairing to account for such high transition temperatures as well as the proximity of superconductivity to antiferromagnetism lead many physicists to believe that magnetism might be the key to understanding the high T_C in these new materials.[2] However, to this day there is still no generally accepted microscopic theory of high temperature superconductivity.

Twenty two years later in 2008 the 1111 iron arsenide family of high temperature superconductors was discovered by H. Hosono's group.[3] Soon thereafter, the discovery of the subclass known as the 122 iron arsenide materials by M. Rotter et al.[4] provided researchers with the opportunity to study large single crystals grown by flux methods. The Fe-As high-temperature superconductors provide researchers with a new system in which superconductivity is found to border on and even possibly coexist[5][6] with a long range antiferromagnetic order of iron moments.

NMR has been shown to be a powerful technique to investigate strongly correlated electron systems such as the ferropnictides. Since NMR is sensitive to both local magnetic fields—produced by SDW ordering of the Fe-As superconductors—and the local electronic structure—via the nuclear quadrupolar coupling to the electric field gradient (EFG)—the technique remains central to the study of Fe-As high temperature superconductors. Recently, ^{75}As NMR has been used to study the effect of

Co and Ni dopants in the 122 system.[7][8] The broadening of the ^{75}As spectra has been interpreted both as (a) the presence of incommensurate AF order, and (b) commensurate AF order with local perturbations in the hyperfine field surrounding the dopants. Here we present spectra of the ^{75}As for $\boldsymbol{H_0} \perp$ c-axis which support the latter interpretation.

2. Nuclear Magnetic Resonance Results

Large single crystals of $Ba(Fe_{1-x}Ni_x)_2As_2$ were grown in Fe-As flux according to [6], [9], and [10]. The plate-like crystals were aligned with the external field perpendicular to the c-axis of the crystal inside of the NMR coil. Field swept spectra were obtained at T = 20K and f = 48.28 MHz by integrating the area under the spin echo as a function of applied field utilizing the same apparatus as [8]. ^{75}As has nuclear spin I=3/2 and the resonance condition is governed by the nuclear spin Hamiltonian:

$$\mathrm{H} = \gamma\hbar\hat{\vec{I}} \cdot \vec{H}_0 + \frac{h\nu_c}{6}\left[3\hat{I}_c^2 - \hat{I} - \eta\left(\hat{I}_a^2 - \hat{I}_b^2\right)\right] + \mathrm{H}_{hf} \qquad (1),$$

where γ = 0.7292 kHz/G is the gyromagnetic ratio; H_0 is the applied field; $\hat{I}_\alpha$ are the nuclear spin operators with respect to the principal axes of the EFG; $\upsilon_c = 3eQV_{cc}/2I(2I\text{-}1)h \cong 2.5$MHz, with e the proton charge, Q the nuclear quadrupole moment, V_{cc} the component of the EFG ǁ c-axis, I the nuclear spin, and h planck's constant; η is the asymmetry parameter of the EFG tensor and H_{hf} is the hyperfine Hamiltonian, given by

$$\mathrm{H}_{hf} = \gamma\hbar\hat{\vec{I}} \cdot \sum_{i\in nn} \vec{B}_i \cdot \vec{S}(\vec{r}_i) \qquad (2),$$

where $\boldsymbol{B_i}$ is the hyperfine coupling tensor and $\boldsymbol{S(r_i)}$ is summed over the four nearest neighbor Fe spins.

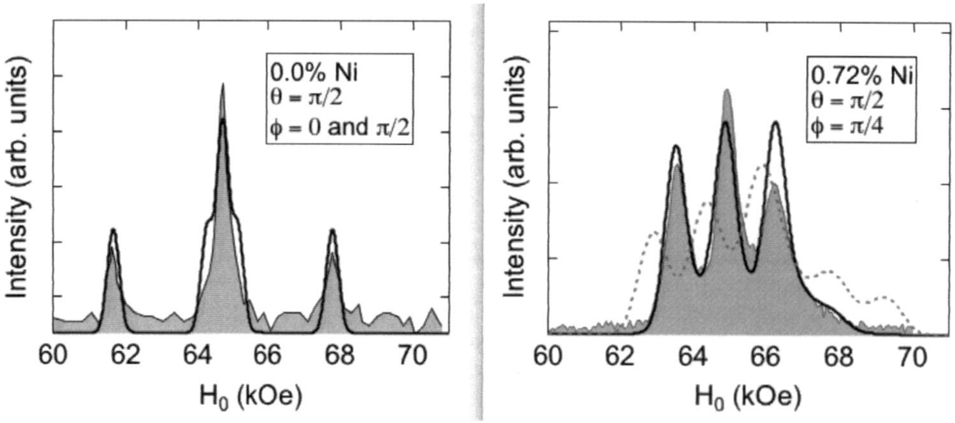

Figure 1. Field-swept spectra of $Ba(Fe_{1-x}Ni_x)_2As_2$ with $H_0 \perp$ c-axis for x = 0 (grey, left) and x = 0.0072 (blue, right). Angles θ and ϕ refer to the crystal orientation with respect to $\boldsymbol{H_0}$ used to calculated the computer simulations (solid black lines). Grey dashed lines refer to incommensurate calculations. The x = 0 data is reproduced from [11].

Below the temperatures $T_S(x)$ and $T_N(x)$, $Ba(Fe_{1-x}Ni_x)_2As_2$ exists in the orthorhombic *Fmmm* crystal structure and a SDW magnetically ordered state respectively.[9][10] In this magnetically ordered state the internal hyperfine field modifies the resonant conditions via H_{hf}. The observed field-swept spectra are denoted by the filled colored lines in Figure 1. The first plot with x = 0 is reproduced from Kitagawa et al.[11] for purposes of comparison to the parent compound. The spectra broaden rapidly as function of doping and it is clear that the crystals become twinned below $T_S(x)$ as described in [12].

3. Computer Simulations

To better understand the subtleties of these field-swept NMR spectra, computer simulations were carried out to calculate resonant field histograms for both the incommensurate and local impurity models. The hyperfine coupling tensor elements were drawn from [11] and combined with the commensurate ordering wave vector Q = (π/a,0,0)[13] to calculate the internal field H_{int} at the ^{75}As site. The magnetic order was then modified by Gaussian suppression around randomly placed impurities of appropriate numbers on a 100 x 100 lattice utilizing the same parameters as [8]. The difference in the resulting magnetic order as well as the in-plane components of the internal hyperfine field near a Nickel dopant are represented in Figure 2.

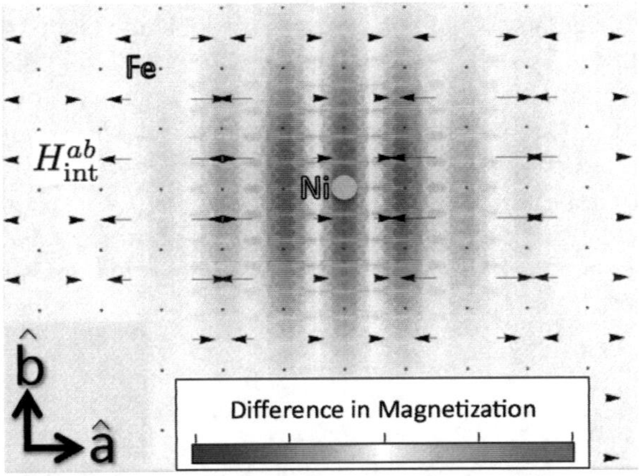

Figure 2. The local perturbation to the SDW order surrounding an impurity is represented by a continuous Blue-to-Red color scale. The in-plane components of the hyperfine field are denoted by black arrows. The dopant influences the magnetic order out to multiple lattice constants.

After calculating the internal field as a function of position, we determine the resonance field by solving Eq. 1 using first order perturbation theory:

$$\gamma\left|\vec{H}_0 + \vec{H}_{\text{int}}\right| = f - p\frac{\nu_c}{2}\left(3\cos^2\theta - 1 + \eta\sin^2\theta\cos(2\phi)\right), \qquad (3)$$

where θ and ϕ are the angles with respect to the total field $\boldsymbol{H}_{tot} = \boldsymbol{H}_0 + \boldsymbol{H}_{int}$. The resulting histograms are represented by the black curves in Figure 1. It is clear that the internal hyperfine field begins to be canted into the plane around a dopant atom and only points along the [100] (a-axis). As a result the orientation of the crystal with respect to the external field will modify the field swept spectra due to both the Zeeman interaction and the quadrupolar interaction. We also show the simulated spectrum (dashed grey lines in Figure 1.) for an incommensurate magnetization **S**(**r**) as described in [7]. Clearly, the local perturbation picture fits the data best.

4. Conclusion

In summation, the ^{75}As NMR field swept spectrum was observed in the SDW state of $Ba(Fe_{1-x}Ni_x)_2As_2$ for x = 0.0072 and compared to the parent compound[11] with the external field applied along the [110] direction and the [100]+[010] respectively. Crystal twinning results in this combination of applied field directions.[12] The spectra broaden rapidly as a function of Nickel substitution as the internal hyperfine field is canted along the [100] direction. Computer simulations within an impurity model based on Gaussian suppression of the local moments are able to reproduce the field swept spectra. The simulations are sensitive to the applied field direction with respect to the magnetic order, and therefore, the crystal axes.

Acknowledgements:

The authors thank the Yukawa Institute for Theoretical Physics at Kyoto University. Discussions during the YITP workshop YITP-W-10-12 on "International and Interdisciplinary Workshop on Novel Phenomena in Integrated Complex Sciences: from Non-living to Living Systems" were useful to complete this work. This workshop was supported in part by the Grant-in-Aid for the Global COE Program "The Next Generation of Physics, Spun from Universality and Emergence" from the Ministry of Education, Culture, Sports, Science and Technology (MEXT) of Japan.

This work was supported by the U.S. Department of Energy, Office of Basic Energy Science, and Division of Materials Sciences and Engineering. The research was performed at Ames Laboratory. Ames Laboratory is operated for the U.S. Department of Energy by Iowa State University under Contract No. DE-AC02-07CH11358.

The authors would also like to thank the conference organizers as well as ICAM/I2CAM for travel funding. A. P. Dioguardi would especially like to thank his advisor and mentor Nicholas J. Curro for sharing his knowledge of NMR as well as his contagious excitement for learning.

References:

[1] Bednorz J G and Müller K A, 1986 *Z. Phys. B* **64**, 189-193
[2] Monthoux P, Pines D and Lonzarich G G 2007 *Nature Reviews* **450** 1177
[3] Kamihara Y, Watanabe T, Hirano M and Hosono H 2008 *J. Am. Chem. Soc.* **130** 3296
[4] Rotter M, Tegel M and Johrendt D 2008 *Phys. Rev. Lett.* **101** 107006
[5] Fernandes R M et al. 2010 *Phys. Rev. B* **81** 140501(R)
[6] Ni N et al. 2008 *Phys. Rev. B* **78** 214515
[7] Laplace Y, Bobroff J, Rullier-Albenque F, Colson D, and Forget A 2009 *Phys. Rev. B* **80** 140501
[8] Dioguardi A P, apRoberts-Warren N, Shockley A C, Bud'ko S L, Ni N, Canfield P C and Curro N J 2010 *Phys. Rev. B* **82** 140411(R)
[9] Canfield P C and Bud'ko S L 2010 *Annu. Rev. Condens. Matter Phys.* **1** 27
[10] Ni N, Thaler A, Yan J Q, Kracher A, Colombier E, Bud'ko S L, Canfield P C, and Hannahs S T 2010 *Phys. Rev. B* **82** 024519
[11] Kitagawa K, Katayama N, Ohgushi K, Yoshida M, and Takigawa M 2008 *J. Phys. Soc. Jpn.* **77** 114709
[12] Tanatar M A, Kreyssig A, Nandi S, Ni N, Bud'ko S L, Canfield P C, Goldman A I and Prozorov R 2009 *Phys. Rev. B* **79** 180508(R)
[13] Goldman A I et al. 2009 *Phys. Rev. B* **79** 024513

Magnetic properties of the novel layered compounds $\boldsymbol{RE}_3$Ru$_4$Al$_{12}$ (RE = La-Nd)

Wanyin Ge, Hiroto Ohta, Chishiro Michioka, Kazuyoshi Yoshimura

Department of Chemistry, Graduate School of Science, Kyoto University, Kyoto 606-8502, Japan

E-mail: wanyinge@kuchem.kyoto-u.ac.jp
kyhv@kuchem.kyoto-u.ac.jp

Abstract. We successfully synthesized a new ternary compound $La_3Ru_4Al_{12}$ using an arc-melting technique and systematically measured magnetization of a series of ternary compounds $RE_3Ru_4Al_{12}$ (RE = La-Nd). X-ray diffraction results demonstrated that all $RE_3Ru_4Al_{12}$ possess the same crystal structure with space group $P6_3/mmc$. $La_3Ru_4Al_{12}$ and $Ce_3Ru_4Al_{12}$ showed paramagnetic behaviors in magnetization, while $Pr_3Ru_4Al_{12}$ and $Nd_3Ru_4Al_{12}$ showed a ferromagnetic ordering at about 40 K. Their Curie temperatures T_C were carefully estimated from so-called Arrott plots.

1. Introduction

In the strongly correlated system, layered compounds are one of the stage where exotic phenomena such as high-T_c superconductivity realize. Recently, a kind of layered compounds $RE_3Ru_4Al_{12}$ gave rise to our extensive attention. In 2002, Niermann *et al.* had synthesized a series of $RE_3Ru_4Al_{12}$ where RE =Pr-Tm except for Pm [1]. Their results revealed that these compounds have $Gd_3Ru_4Al_{12}$-type structure with space group $P6_3/mmc$ (194). In 2004, Bukhan'ko *et al.* successfully synthesized $Ce_3Ru_4Al_{12}$ and showed that it also possesses the $Gd_3Ru_4Al_{12}$-type structure [2]. Although these research groups have contributed their endeavors to syntheses of $RE_3Ru_4Al_{12}$, there is no report of magnetic properties of $RE_3Ru_4Al_{12}$. In order to understand physical property of $RE_3Ru_4Al_{12}$, we believe it is of importance to study magnetic properties of $RE_3Ru_4Al_{12}$. Therefore, in this paper, we present the syntheses and magnetic properties of $RE_3Ru_4Al_{12}$ compounds.

2. Experiments and characterization

An argon arc-melting technique was employed to synthesize intermetallic compounds $RE_3Ru_4Al_{12}$ (RE=La-Nd). High-purity powders of RE (RE=La-Nd), Ru, Al (purity: 99.99 %) were mixed in a stoichiometric ratio. After a grind of the mixed powders in agate mortar, the mixture was cold-pressed to pellets and was melted in an argon arc-melting furnace. The pellets were melted at least three times for each side to make sure the homogeneity. The obtained button-like samples were annealed at 800°C in a high vacuumed quartz tube for one week and subsequently were quenched with the mixture of ice and water.

The quenched samples were smashed into fine powder for X-ray diffraction (XRD) measurements. XRD data were collected with the θ-2θ scanning scope from 2θ=10° to 80°

with a step of 0.02° using Cu K_α radiation. The Rietveld refinement was performed by GSAS soft package [5]. Magnetization of the powder samples was measured using a superconducting quantum interference device magnetometer installed in Research Center for Low Temperature and Materials Sciences, Kyoto University in the temperature range between 2 and 300 K.

3. Results and discussions

Figure 1 shows the powder XRD patterns of $RE_3Ru_4Al_{12}$ (RE=La-Nd) and the simulated pattern for $Gd_3Ru_4Al_{12}$ with space group $P6_3/mmc$. $La_3Ru_4Al_{12}$ shows a similar diffraction pattern to the other compounds, which indicate that $La_3Ru_4Al_{12}$ is successfully synthesized and has the same $Gd_3Ru_4Al_{12}$-type structure. From the XRD results all the compounds were confirmed to be in a single phase of $RE_3Ru_4Al_{12}$ without any impurities.

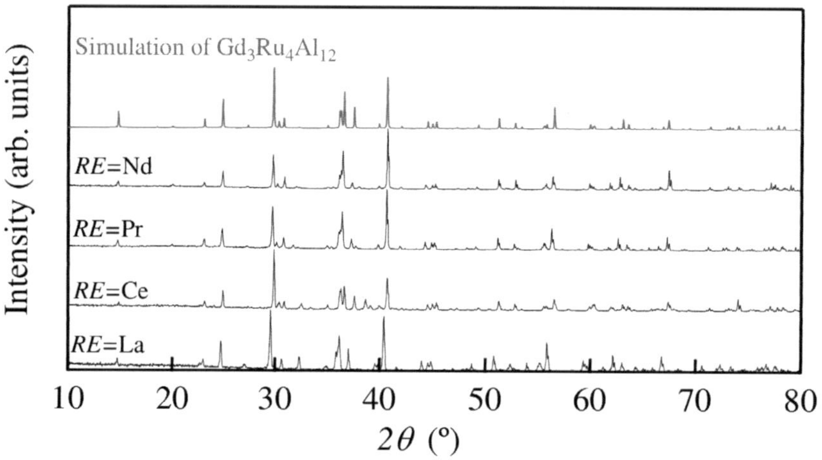

Figure 1. Powder XRD patterns measured for $RE_3Ru_4Al_{12}$ (RE=La-Nd) and the simulated one for $Gd_3Ru_4Al_{12}$ with space group $P6_3/mmc$ (red line).

Table 1. Lattice parameters of $RE_3Ru_4Al_{12}$ (RE=La-Nd) and reported values in Refs. [1, 2].

Parameter	a (Å)	c (Å)	V (Å^3)
$La_3Ru_4Al_{12}$: this work	8.9254	9.7015	669.31
No Ref.	-	-	-
$Ce_3Ru_4Al_{12}$: this work	8.8646	9.5744	651.57
Ref. [2]	8.8705	9.5536	650.90
$Pr_3Ru_4Al_{12}$: this work	8.8824	9.6508	659.41
Ref. [1]	8.8691	9.6331	656.20
$Nd_3Ru_4Al_{12}$: this work	8.8552	9.6202	653.35
Ref. [1]	8.8491	9.611	651.80

As the results of the Rietveld refinement, all the compounds were confirmed to have the hexagonal structure with $P6_3/mmc$. The estimated lattice parameters for a, c and unit cell volume V are listed in Table 1. Our results agree with the published data for the compounds RE= Ce, Pr and Nd [1]. The lattice constants a and c of $RE_3Ru_4Al_{12}$ (RE=La-Nd)

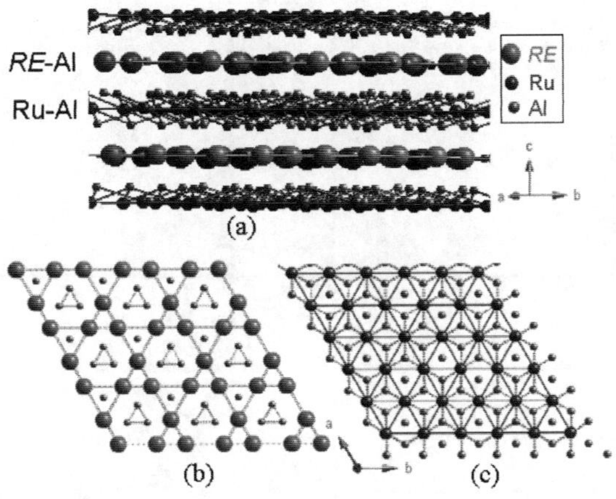

Figure 2. (a): Crystal structure of $RE_3Ru_4Al_{12}$; (b) and (c): the projection view of the "RE-Al" and the "Ru-Al" layers along its c-axis.

monotonically decrease from La to Nd except for Ce. Only for $Ce_3Ru_4Al_{12}$, lattice parameters are smaller than expected values from the ion radii of RE^{3+}, indicating that Ce ion is not in a trivalent state.

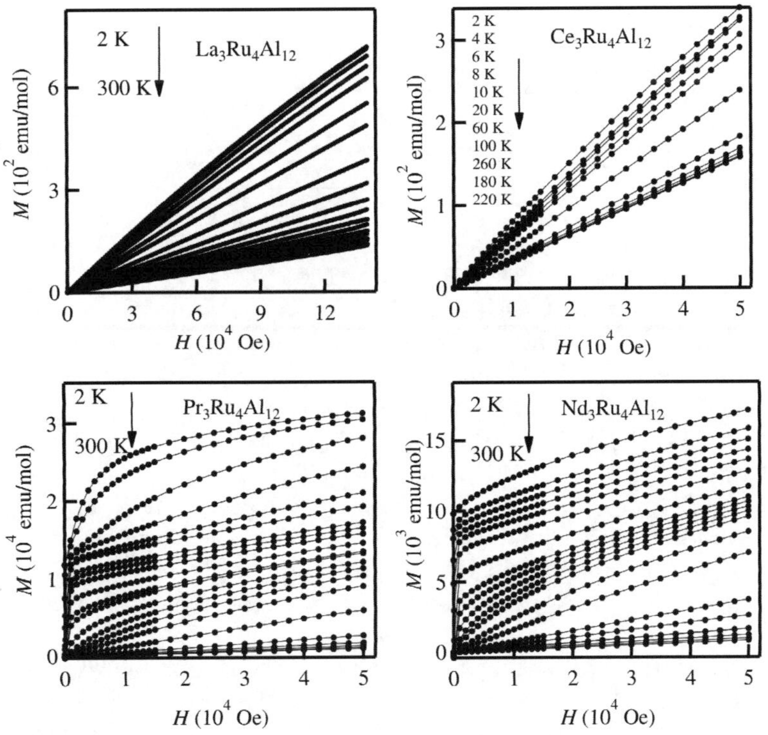

Figure 3. M versus H for $RE_3Ru_4Al_{12}$ (RE=La-Nd) at various temperatures.

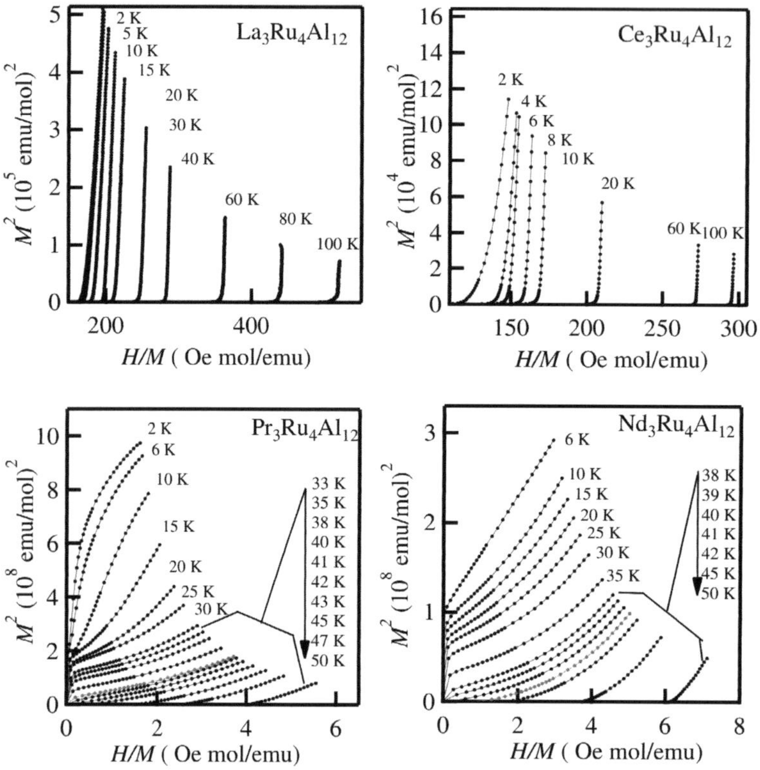

Figure 4. Arrott plots of magnetization, M^2 versus H/M, for $RE_3Ru_4Al_{12}$ (RE=La, Ce, Pr and Nd).

The structure of $RE_3Ru_4Al_{12}$ as shown in Fig. 2(a), is made up of the "RE-Al" and the "Ru-Al" layers, which are stacking alternately along the c-axis. In the "RE-Al" layer shown in Fig. 2(b), RE atoms form a distorted-Kagomé lattice, and Al atoms occupy two different sites in this plane. Three Al atoms form a triangle with an Al atom in the center of RE triangle. For Ru_4Al_8 unit shown in Fig. 2(c), Ru atoms form a regular triangular lattice and Al atoms are located a little bit out-of-layer as shown in Fig. 2(a).

Figure 3 shows magnetic field (H) dependence of magnetization (M) of the samples at various temperatures. For $La_3Ru_4Al_{12}$ and $Ce_3Ru_4Al_{12}$, both show the linear relation between M and H, indicating that both compounds are in a paramagnetic state. It should be noted that Ru atoms in $RE_3Ru_4Al_{12}$ can have magnetic moments because of the Curie-Weiss like paramagnetic behavior in $La_3Ru_4Al_{12}$. For $Pr_3Ru_4Al_{12}$ and $Nd_3Ru_4Al_{12}$, on the other hand, both compounds show a typical ferromagnetic behavior; in the low temperature region, M's of both compounds show a nearly saturation behavior.

As the result of the Landau expansion of free energy of the system by $M(H,T)$ in the thermal equilibrium state, M^2 will display the linear relation against H/M in a ferromagnetic case when the higher terms of Landau expansion are neglected. The equation can be expressed as follows.

$$M(H,T)^2 = a + b \cdot \frac{H}{M(H,T)}, \tag{1}$$

where $a \propto (1 - T/T_C)$ and T_C is the Curie temperature. If one plots $M^2(H,T)$ against $H/M(H,T)$, the line will cross the origin when $T = T_C$. Therefore, T_C can be retrieved from the equation (1) which usually named as Arrott plot [6]. Arrott plots of $RE_3Ru_4Al_{12}$ (RE=La-Nd) from 2 to 300 K are shown in Fig. 4. For $La_3Ru_4Al_{12}$ and $Ce_3Ru_4Al_{12}$, both show almost the linear relation between M^2 and H/M. There is no curve passing through positive interception and the origin of the coordination, suggesting that $La_3Ru_4Al_{12}$ and $Ce_3Ru_4Al_{12}$ do not show any ferromagnetic orderings at least down to 2 K. In the cases of $Pr_3Ru_4Al_{12}$ and $Nd_3Ru_4Al_{12}$, on the other hand, curves pass the positive interception below 40 K and there are exact curves that pass through the origin of the coordination. Therefore, we can estimate the values of T_C as 41 and 40 K for $Pr_3Ru_4Al_{12}$ and $Nd_3Ru_4Al_{12}$, respectively. It should be pointed out that M^2 vs. H/M does not obey the strict linear relation.

4. Conclusions

In summary, we successfully synthesized series of $RE_3Ru_4Al_{12}$ with RE = La-Nd using the arc-melting technique. For $Pr_3Ru_4Al_{12}$ and $Nd_3Ru_4Al_{12}$, a huge value of magnetization and saturation in it against magnetic fields suggest ferromagnetic orders. Based on the Arrott plots, T_C of $Pr_3Ru_4Al_{12}$ and $Nd_3Ru_4Al_{12}$ are estimated as 41 and 40 K, respectively. We found that in $RE_3Ru_4Al_{12}$, the electronic state changes drastically for RE between Pr and Nd, resulting in the ferromagnetic ordering in RE= Pr, Nd.

5. Acknowledgements

This workshop was supported in part by the Grant-in-Aid for the Global COE Program "the Next Generation of Physics, Spun from Universality and Emergence" from the Ministry of Education, Culture, Sports, Science and Technology (MEXT) of Japan. The authors thank the Yukawa Institute for Theoretical Physics at Kyoto University, where this work was initiated [completed] during the YITP-W-10-12 on "International and Interdisciplinary Workshop on Novel Phenomena in Integrated Complex Sciences: from Non-living to Living Systems". This work is also supported by Grants-in-Aid for Scientific Research from the Japan Society for Promotion of Science (19350030, 22350029).

References

[1] Niermann J, Jeitschko W 2002 *Z. Anorg Allg. Chem.* **628** 2549
[2] Bukhan'ko N G, Tursina A I, Malyshev S V, Gribanov A V, Seropegin Y D and and Bodak O I 2004 *J. Alloys Compd.* **367** 149
[3] Cordier G, Dörsam G and Kniep R 1988 *J. Magn. Magn. Mater.* **76-77** 653
[4] Gladyshevskii R E, Strusievicz O R, Cenzual K and Parthé E 1993 *Acta Crystallogr B* **49** 474
[5] Toby B, 2001 *J. Appl. Crystallogr.* **34**, 210
[6] Arrott A 1957 *Phys. Rev.* **108** 1394

Journal of Physics: Conference Series **344** (2012) 012024 doi:10.1088/1742-6596/344/1/012024

Single crystal growth and magnetic property of $NH_4CuPO_4 \cdot H_2O$

K. Nawa[1], C. Michioka[1], K. Yoshimura[1], A. Matsuo[2] and K. Kindo[2]

[1] Department of Chemistry, Graduate School of Science, Kyoto University, Kitashirakawa Oiwake-cho, Sakyo-ku, Kyoto, Japan

[2] The Institute for Solid State Physics, The University of Tokyo, Kashiwanoha, Kashiwa, Japan

E-mail: `knawa@kuchem.kyoto-u.ac.jp`, `kyhv@kuchem.kyoto-u.ac.jp`

Abstract. We synthesized single crystals of $NH_4CuPO_4 \cdot H_2O$ by a chemical reaction in aqueous solutions. Magnetization and specific heat are well reproduced by a dimer model, and $NH_4CuPO_4 \cdot H_2O$ does not show any field-induced magnetic phase transitions. The interdimer interactions would be too weak to observe field-induced phase transitions.

1. Introduction

An antiferromagnetically coupled spin dimer system without interdimer interactions has a singlet ground state. In spin dimers with interdimer interactions, an antiferromagnetic long range ordered (AFLRO) state could be induced by magnetic fields. The field-induced magnetic phase transition can be regarded as a magnetic ordering of their transverse spin components by using the mean field approximation [1]. From the power-law behaviour of the phase boundary in $TlCuCl_3$, the AFLRO state is claimed to be better described by a Bose-Einstein condensation (BEC) of excited magnons [2]. It is important to explore spin-gap compounds which show the AFLRO phase to investigate the nature of the field-induced magnetic phase transition.

In this paper, we report an exploration of an AFLRO phase in a layered compound $NH_4CuPO_4 \cdot H_2O$, which includes copper phosphate layers separated by NH^{4+} cations [3, 4, 5]. It crystallizes in a monoclinic structure (space group $P2_1/a$) which is related to that of the dittmarite ($NH_4MgPO_4 \cdot H_2O$) family. A pair of CuO_5 distorted square pyramids in the layer shares an edge and forms a Cu_2O_8 dimer, which is bridged by phosphate tetrahedra. The arrangement of Cu^{2+} ions in a layer and exchange interactions J_1 - J_4 are shown in Figure 1(a). The magnetic susceptibility agrees with an isolated dimer model of $S = 1/2$ with a spin gap of 9.6 K [3]. The quantitative mapping analysis based on first principle electronic structure calculations predicted that magnetic properties of $NH_4CuPO_4 \cdot H_2O$ should be described by an antiferromagnetic

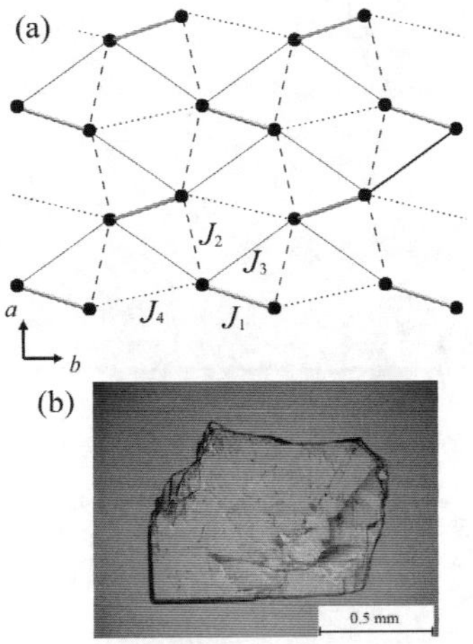

Figure 1. (a) Equivalent spin model of $NH_4CuPO_4 \cdot H_2O$. Solid circles express Cu^{2+} ions and lines represent the spin interactions which is equivalent with that in reference [4]. (b) Viewgraph of a single crystal.

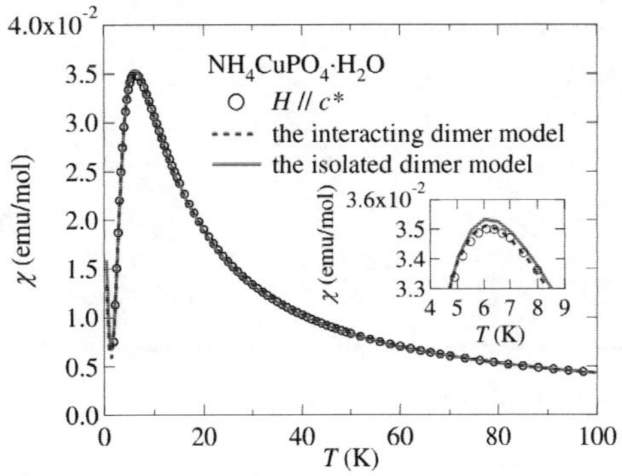

Figure 2. Temperature dependence of magnetic susceptibility with a field applied to the c^*-axis. Dashed and solid curves are fitted curves with the interacting dimer and the isolated dimer models, respectively. Inset shows magnetic susceptibility between 4 and 9 K.

- ferromagnetic (AF-F) alternating chain model [4]. That is, antiferromagnetic interactions J_1 and ferromagnetic interactions J_4 form the alternating chains along the b-axis, which couple each other by weak interactions of J_2 and J_3.

2. Experimental

$NH_4CuPO_4 \cdot H_2O$ was synthesized by a chemical reaction in aqueous solutions [5]. Aquenous solutions were prepared by dissolving 0.6 g of $CuCl_2 \cdot H_2O$ and 4.0 g of $(NH_4)_2HPO_4$ in 10 mL and 100 mL of distilled water, respectively. The solutions were mixed and then 1 mL of 1 mol/L HCl was added. After preserving the mixed solution for two months, tiny greenish blue crystals were obtained. The crystals were separated by filtering and washed by distilled water. Figure 1(b) shows a typical single crystal with a size of $1 \times 2 \times 0.1$ mm^3. Crystals were confirmed to be in a single phase by powder XRD measurements. The crystal axes were determined by transmission photographs. Magnetic susceptibility of a single crystal was measured by a superconducting quantum interference device magnetometer. High-field-magnetization was measured for a polycrystalline sample up to 20 T with a pulse magnet at Ultra High Magnetic Field Laboratory, Institute for Solid State Physics, the University of Tokyo. Specific heat of a polycrystalline sample was measured by a relaxation method under a magnetic field up to 14 T.

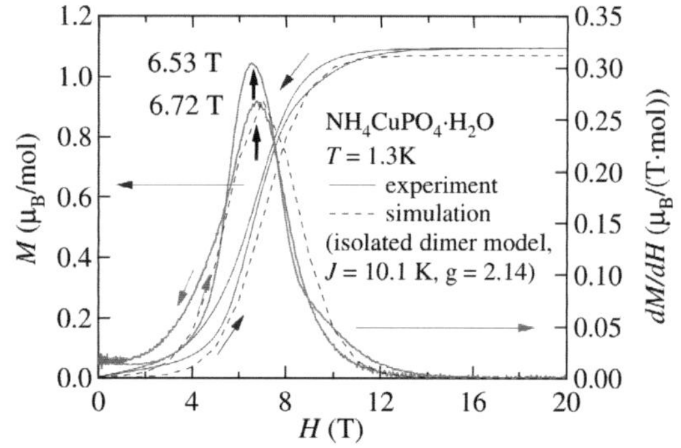

Figure 3. High-field magnetization and its first field-derivative dM/dH of a polycrystalline sample. Solid and dashed curves stand for the experimental data and a simulation of the isolated dimer model, respectively.

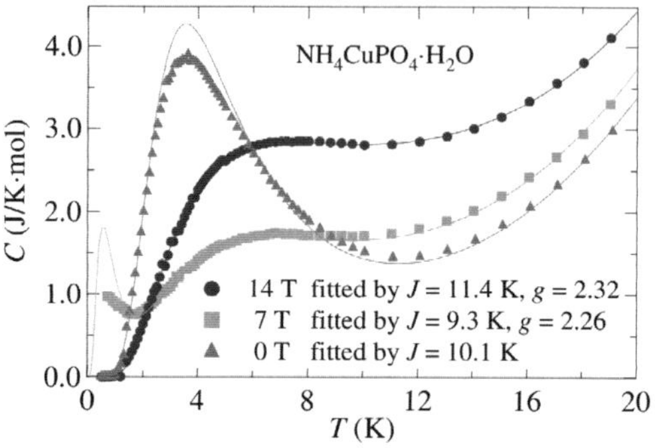

Figure 4. Specific heat of a polycrystalline sample under applied fields of 0, 7 and 14 T. Solid curves show fitted curves with the isolated dimer model.

3. Results and discussion

Figure 2 shows the magnetic susceptibility with a field applied to the c^*-axis. The susceptibility was analyzed by both an isolated spin dimer model [6] and an interacting dimer model. The Curie term was added to the fitting functions to describe a contribution of magnetic impurities, which should be quite small. The fitting function of the isolated dimer model is expressed as

$$\chi_{\text{isodimer}} = \chi_{\text{iso}} + \frac{C}{T} + \chi_0, \tag{1}$$

where χ_{iso}, C, χ_0 are the spin part of the isolated dimers, the Curie constant of the Curie term and the temperature independent term, respectively. In the dimer model with interdimer interactions, interdimer interactions are introduced as effective fields. The fitting function is described by

$$\chi_{\text{intdimer}} = \frac{\chi_{\text{iso}}}{1 + \frac{J'}{N(g\mu_B)^2}\chi_{\text{iso}}} + \frac{C}{T} + \chi_0, \text{ with } J' = \sum_i \frac{z_i}{2} J_i, \tag{2}$$

where z_i is the number of nearest neighbors connected by the spin interactions J_i. Compared with the spin model shown in Figure 1(a), the parameters J and J' are expected to correspond to $J = J_1$ and $J' = 2J_2 + 2J_3 + J_4$, respectively. Values of the fitting parameters are listed in Table 1. The interacting dimer model explains the experimental data better than the isolated dimer one. A positive value of J' indicates that interdimer interactions are antiferromagnetic in this system, which is in contrast to the alternating chain model with ferromagnetic interdimer interactions suggested by quantitative mapping analysis [4].

The high-field magnetization and its first field-derivative are shown in Figure 3. No clear anomaly originated in the phase transition was observed in the magnetization curve. Eventually, the magnetization curve is explained by the isolated dimer model with the intradimer interaction of $J = 10.1$ K.

Temperature dependence of the specific heat is shown in Figure 4. Some Schottky peaks were observed above 0.5 K, reflecting energy gaps in the dimers. The specific heat was fitted to the superposition of isolated dimers and a phonon contribution as the Debye model. An intradimer interaction J, a g-factor g and a Debye temperature θ (in zero field, only J and θ) were used for the fitting. The fitting curves and parameters are shown in Figure 4. An evaluation of a phonon term is confirmed to be correct by agreement of the Debye temperatures in different fields, all of which were within 170 ± 2 K. The J is estimated as between 9.3 and 11.4 K, which is consistent with the magnetization measurements. A slight difference between the parameters obtained from the magnetization and the specific heat measurements would be attributed to anisotropy of g-tensor or to the presence of interdimer interactions.

Let us allow to discuss the condition for the appearence of an AFLRO phase by a magnetic field. From the equivalent spin model in Figure 1, its spin Hamiltonian can be written as

$$H = H_0 + H' \tag{3}$$

$$H_0 = J_1 \sum_i \mathbf{s}_{i1} \cdot \mathbf{s}_{i2} - g\mu_B H \cdot \sum_i (\mathbf{s}_{i1} + \mathbf{s}_{i2}) \tag{4}$$

$$H' = J_2 \sum_{\langle i,j \rangle_2} \mathbf{s}_{i1} \cdot \mathbf{s}_{j1} + J_3 \sum_{\langle i,j \rangle_3} \mathbf{s}_{i1} \cdot \mathbf{s}_{j2} + J_4 \sum_{\langle i,j \rangle_4} \mathbf{s}_{i1} \cdot \mathbf{s}_{j2}, \tag{5}$$

where $i1$ and $i2$ represent the i-th pair of spins, and $\langle i, j \rangle_n$ represents the exchange path described by the interdimer interaction parameter $J_n (n = 2, 3, 4)$. In the classical model with above interactions, a symmetric bell-shape phase boundary of the AFLRO phase is expected. If we assume that spin frustration is absent, i.e. $J_3 = 0$ and $J_4 > 0$, a lower critical field H_{c1}, an upper critical field H_{c2} and a maximum transition temperature $T_{\max}$ can be estimated as

$$H_{c1} = \frac{1}{g\mu_B} \left(J - \frac{J'}{2} \right) = 5.8 \text{ T}, \tag{6}$$

$$H_{c2} = \frac{1}{g\mu_B} (J + J') = 8.8 \text{ T}, \tag{7}$$

$$T_{\max} = \frac{J'}{4k_B} = 0.72 \text{ K}. \tag{8}$$

Table 1. Parameters obtained from fitting the magnetic susceptibility.

Parameters	Isolated dimers	Interacting dimers
J (K)	10.1 ± 0.0	10.0 ± 0.0
J' (K)		2.88 ± 0.23
g	2.14 ± 0.00	2.19 ± 0.00
C (emu K/mol)	$(7.96 \pm 0.33) \times 10^{-3}$	$(6.80 \pm 0.25) \times 10^{-3}$
χ_0 (emu/mol)	$(7.45 \pm 2.61) \times 10^{-5}$	$-(12.1 \pm 2.36) \times 10^{-5}$

The above estimation indicates that the transition temperature should be very low in $NH_4Cu_2PO_4 \cdot H_2O$ because of weak interdimer interactions. Experimentally, we cannot observe any AFLRO phase transition down to 0.5 K and up to 14 T in $NH_4Cu_2PO_4 \cdot H_2O$. The weak interdimer interactions J' would allow the occurence of a long-range ordering only at very low temperatures. Explorations of the spin-gap system with strong interdimer interactions are prefered to observe AFLRO phase transitions.

Acknowledgements

This workshop was supported in part by the Giant-in-aid for the Global COE Program "The Next Generation of Physics, Spun from Universality and Emergence" from the Ministy of Education, Culture, Sports, Science and Technology (MEXT) of Japan. This work was supported by a Grant-in-Aid for Scientific Research from the Japan Society for Promotion of Science (19350030, 22350029) and also by Global COE Program of Kyoto University, "International Center for Integrated Research and Advanced Education in Materials Science", from the Ministry of Education, Culture, Sports, Science and Technology of Japan. The authors thank the Yukawa Institute for Theoretical Physics at Kyoto University, where this work was completed during the YITP-W-10-12 on "International and Interdisciplinary Workshop on Novel Phenomena in Integrated Complex Sciences: from Non-Living to Living Systems".

References

[1] Tachiki M and Yamada T 1970 *J. Phys. Soc. Jpn.* **28** 1413.
[2] Nikuni T *et al.* 2000 *Phys. Rev. Lett.* **84** 5868.
[3] Pujana *et al.* 1998 *Chem. Mater* **8** 1055.
[4] Koo H-J *et al.* 2008 *J. Solid State Chem.* **181** 276.
[5] Basset H and Bedwell W L **1933** *J. Chem. Soc.* 854.
[6] Figgis B N and Martin R L **1956** *J. Chem. Soc.* 3837.

Magnetic properties of layered compounds *Ln*CoAsO (*Ln* = lanthanoids) with itinerant-electron ferromagnetism

Hiroto Ohta, Chishiro Michioka and Kazuyoshi Yoshimura
Department of Chemistry, Graduate School of Science, Kyoto University, Kyoto 606-8502, Japan

E-mail: shioshio@kuchem.kyoto-u.ac.jp

Abstract. We synthesized polycrystalline samples of layered compounds *Ln*CoAsO (*Ln* = La - Gd) and measured magnetization of the samples, especially, in the cases of *Ln* = Nd, Sm and Gd, which show ferromagnetic-antiferromagnetic transitions (FAFT). From the results of magnetic measurements, we successfully draw the magnetic phase diagrams of these compounds. We showed that the FAFT temperatures T_N at various magnetic fields in the cases of *Ln* = Nd and Sm well scale to the de Gennes factors of the trivalent ions of Nd and Sm, showing that the Ruderman-Kittel-Kasuya-Yoshida interaction is the origin of the FAFT in these compounds. We also draw the total magnetic phase diagram of *Ln*CoAsO (*Ln* = La - Gd).

1. Introduction

Recent discoveries of superconductivity with high superconducting transition temperature up to 55 K in doped *Ln*FeAsO [1, 2] and in the related compounds[3, 4, 5, 6] have aroused much interest. On the other hand of tremendous number of studies on the Fe-based superconductors, the number of studies on Co-based compounds such as *Ln*CoAsO have been increasing recently. Among them, LaCo*Pn*O (*Pn* = P, As) has been still basic and important since they are suitable to a study on electronic natures of Co*Pn* planes. LaCoAsO and LaCoPO were reported to show a ferromagnetic ordering below the Curie temperature (T_C) equal to about 50 and 60 K, respectively [7, 8, 9]. Magnetic properties of these two compounds can be almost understood within a framework of the ferromagnetic spin-fluctuation theory, and two-dimensional nature of spin fluctuations has also been pointed out [9].

As reported by both Zimmer and Quebe, lanthanum in LaCo*Pn*O can be substituted by other lanthanoid elements [10, 11]. The 4*f*-electrons of Ln^{3+} with a localized character interact with itinerant electrons of Co through the Ruderman-Kittel-Kasuya-Yoshida (RKKY) interaction. As a result, a magnetic structure of ferromagnetically ordered moments of Co is forced to change in the low-temperature region. Especially, in the cases of *Ln* = Nd, Sm and Gd, a ferromagnetic-antiferromagnetic transition (FAFT) was observed below T_C [12, 13, 14, 15, 16]. Recently, we studied the magnetic property of SmCoAsO and showed its detailed magnetic phase diagram [16]. It should be interesting to compare the magnetic phase diagrams of NdCoAsO, SmCoAsO, and GdCoAsO with each other, since it gives a hint for the detailed mechanism of FAFTs in those compounds.

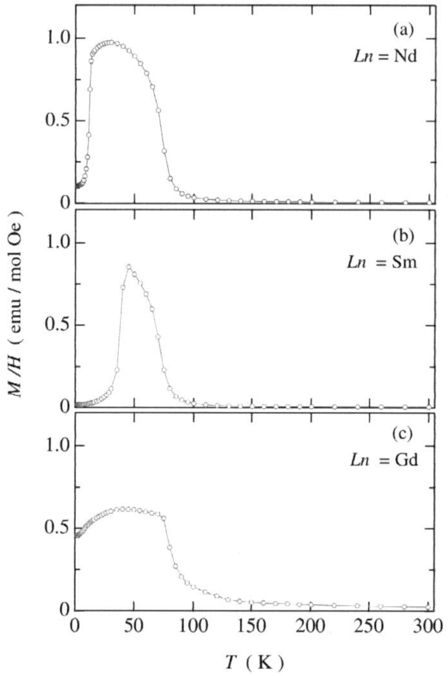

Figure 1. T dependence of M/H of (a) NdCoAsO, (b) SmCoAsO and (c) GdCoAsO measured at H = 1 kOe.

In this report, we showed detailed results of magnetic measurements on polycrystalline samples of LnCoAsO (Ln = Nd, Sm and Gd). From the results of measurements, we draw the magnetic phase diagrams of these three compounds and compared the curvatures of the FAFT line.

2. Experiments

For the synthesis of polycrystalline samples of LnCoAsO, we used powders of Ln (= Nd, Sm and Gd) (purity: 99.9%), As (99.99%) and CoO (99.99 %) as starting materials. Powders of Ln and As were first mixed and sealed in an evacuated silica tube. The sealed tube was carefully fired in a furnace at 550°C for 5 h and then at 800°C for 12 h. The obtained powder of LnAs was mixed with powder of CoO to a stoichiometric ratio and was ground well in hexane to avoid oxidation. The pelletized mixture of LnAs and CoO was sealed in an evacuated silica tube and fired at 1000°C for 12 h.

We measured powder X-ray diffractions (XRDs) of the obtained samples and confirmed them to be in a single phase of LnCoAsO with the space group of $P4/nmm$. We estimated lattice parameters from the powder XRD patterns and confirmed to be the same with the previous report[11]. The detailed XRD results can be seen in our previous report[12]. Magnetization (M) of the obtained samples was measured as a function of temperature (T) and magnetic field (H) by using a superconducting quantum interference device magnetometer installed in Research Center for Low Temperature and Materials Sciences, Kyoto University up to 55 kOe.

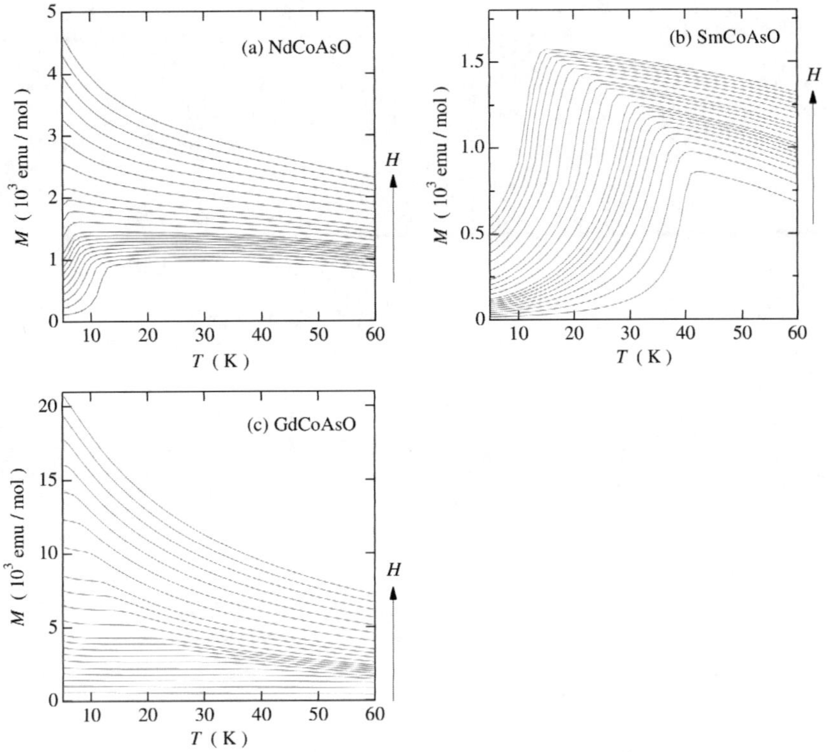

Figure 2. T dependence of M of (a) NdCoAsO, (b) SmCoAsO, and (c) GdCoAsO at $H = 1 \sim 9$ kOe with every 1 kOe, $10 \sim 20$ kOe with every 2.5 kOe and $25 \sim 55$ kOe with every 5 kOe.

3. Results and Discussions

Figure 1 shows T dependence of M/H of (a) NdCoAsO, (b) SmCoAsO and (c) GdCoAsO measured at $H = 1$ kOe. In the high-T region, M of all the samples showed a Curie-Weiss like increase with decreasing T, while in the low-T region M of NdCoAsO and SmCoAsO showed an abrupt decrease around 15 and 40 K, respectively. In the case of GdCoAsO, M showed a kink at about 75 K and showed a broad peak around 30 K. We have reported in the previous paper that these compounds show a ferromagnetic transition at about 75 K and an FAFT at the low-T region[12]. Especially, in the case of GdCoAsO both the ferromagnetic transition and the FAFT simultaneously occur at about 75 K. We have pointed out that because the FAFT temperature (T_N) well scales to the de Gennes factor of Ln^{3+} ions the FAFT corresponds to an antiferromagnetic ordering of magnetic moments of $4f$-electrons of Ln^{3+} ions and ferromagnetic ordered magnetic moments of $3d$-electrons of Co through the RKKY interaction[12]. In the case of GdCoAsO, according to the value of the de Gennes factor, the FAFT must occur around 150 K. However, in such a high-T region, magnetic moments of Co do not order. Therefore, T_N of GdCoAsO is suppressed to 75 K where magnetic moments of Co ions ferromagnetically order.

To study the magnetic phase diagram of these compounds in detail, we precisely measured magnetization around T_N. Figure 2 shows T dependence of M of (a) NdCoAsO, (b) SmCoAsO and (c) GdCoAsO between 5 and 60 K at $H = 1 \sim 9$ kOe with every 1 kOe, $10 \sim 20$ kOe with every 2.5 kOe and $25 \sim 55$ kOe with every 5 kOe. The T_N decreased with the increase of H in the cases of NdCoAsO and SmCoAsO. In the case of GdCoAsO, the kink temperature which also indicates T_N decreased with the increase of H. In both cases of NdCoAsO and GdCoAsO, T_N

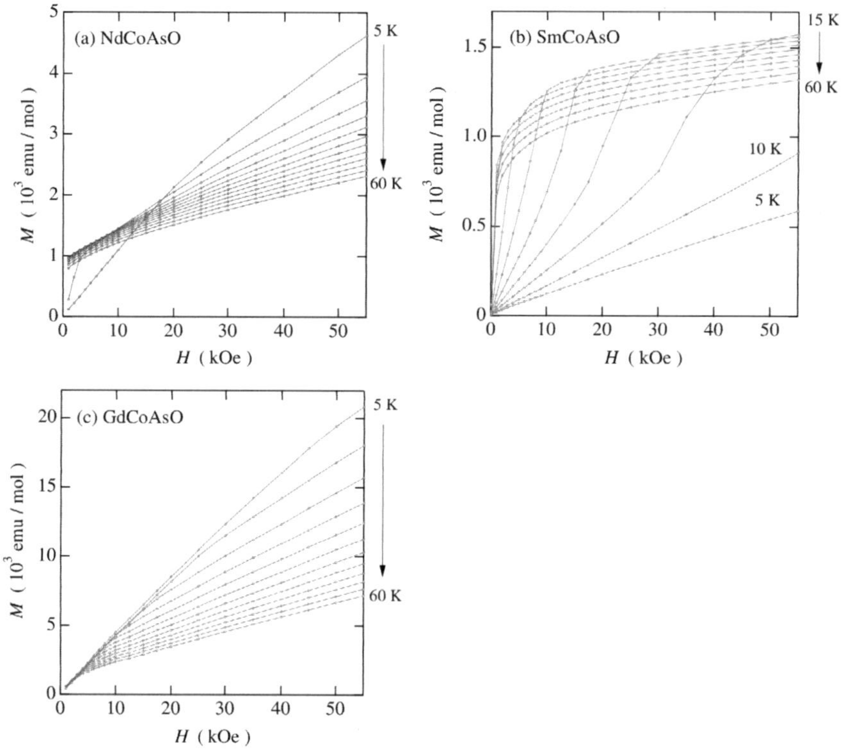

Figure 3. H dependence of M of (a) NdCoAsO, (b) SmCoAsO and (c) GdCoAsO at $T = 5 \sim 60$ K with every 5 K.

is completely suppressed to zero within 55 kOe, while in the case of SmCoAsO higher magnetic field is needed to suppress T_N to zero.

Figure 3 shows H dependence of M at $T = 5 \sim 60$ K with every 5 K. With increasing H, jumps of M indicating the magnetic field induced FAFT were observed in all the samples below T_N. Since the magnetic moments of $4f$-electrons of Sm^{3+} is much smaller than those of other two lanthanoids, the FAFT is much clearer than in the cases of other two compounds. We measured M of SmCoAsO at 4.2 K up to 650 kOe (65 T) using a pulsed magnet and at various T up to 140 kOe (14 T) using a 14 T superconducting magnet, and successfully draw the magnetic phase diagram[16]. Those data will appear in the magnetic phase diagram later.

From the results of T dependence of M of each sample, we successfully draw the magnetic phase diagram of three compounds. Figure 4 (a) shows the H versus T phase diagram of the three compounds. Closed and open circles show data of NdCoAsO and SmCoAsO, respectively. Inset of Fig. 4 (a) shows the data of GdCoAsO. Closed diamond shows the T_C ($\sim$ 75 K) of both NdCoAsO and SmCoAsO. Dashed lines are guides for the eyes. By naturally extrapolating the phase boundary of FAFT to $T = 0$, one can roughly estimate the value of transition field at the ground state (H_N) as 50 kOe for NdCoAsO, 250 kOe for SmCoAsO and 50 kOe for GdCoAsO. The FAFT line of SmCoAsO is quite similar to that of NdCoAsO in its shape. On the other hand, the FAFT line in the case of GdCoAsO is different from those of other two compounds. We showed the value of T_N at zero field except for the case of Ln = Gd scaling to the de Gennes factor (dGf) of Ln^{3+}[12]. It seems interesting to compare the H vs T phase diagrams in a normalized form by the de Gennes factor. Figure 4 (b) shows the H/dGf versus T/dGf phase

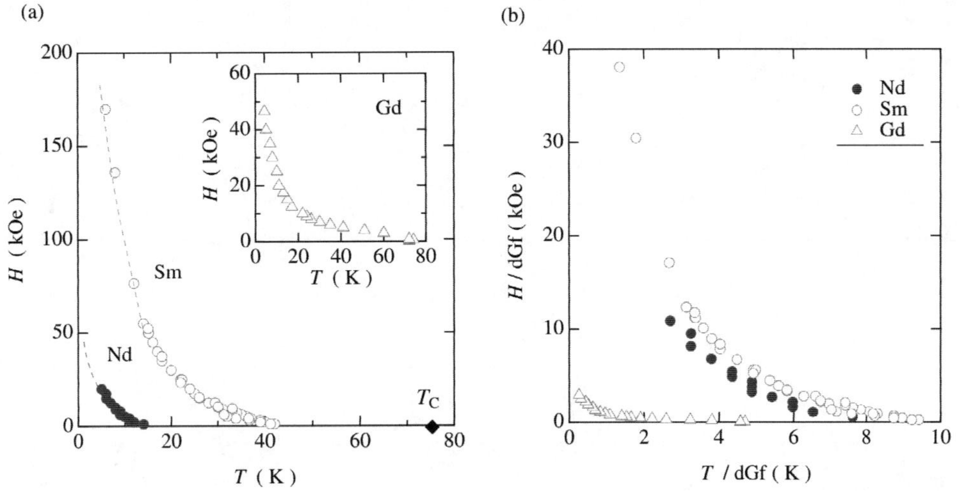

Figure 4. (a) H versus T phase diagram of LnCoAsO. Closed and open circles show the data of NdCoAsO and SmCoAsO, respectively. Closed diamond shows Curie temperature of both samples ($\sim$ 75 K). Dashed lines are guides for the eyes. Inset: H-T phase diagram of GdCoAsO. (b) Modified magnetic (H/dGf vs T/dGf) phase diagram of LnCoAsO. Closed and open circles show the data of NdCoAsO and SmCoAsO, respectively and open triangles show those of GdCoAsO. The envelopes of the Nd and Sm cases almost overlap with each other, while in the case of Gd the envelope does not scale to those of the other two samples.

diagram of the compounds. Here, the values of dGf of Nd^{3+}, Sm^{3+}, and Gd^{3+} are 81/44, 125/28, and 63/4, respectively. The FAFT lines of NdCoAsO and SmCoAsO almost overlap with each other, showing that the origins of FAFT in both cases are really the same with each other. On the other hand, the FAFT line of GdCoAsO is solely different from those of other two compounds, indicating some different mechanism induces the antiferromagnetic order in GdCoAsO. Since magnetic moments of Gd^{3+} are large and have no orbital part, antiferromagnetically ordered moments may easily flop in relatively weak magnetic field. Finally, we summarize the data of magnetic measurements as the total magnetic phase diagram of LnCoAsO (Ln = La - Gd) in Fig. 5. Dashed lines are guides for the eyes. PM, FM, and AFM denote a paramagnetic phase, a ferromagnetic phase and an antiferromagnetic phase, respectively. The dGf lines correspond to the de Gennes factor of Ln^{3+} ions which are scaled to T_N or H_N of NdCoAsO and SmCoAsO.

4. Summary

We measured magnetizations of polycrystalline samples of LnCoAsO with Ln = Nd, Sm and Gd as functions of temperature and magnetic field. We successfully observed the magnetic field dependence of the ferromagnetic-antiferromagnetic transition (FAFT) up to 55 kOe for each sample, and draw the magnetic phase diagram of the samples. The FAFT lines of NdCoAsO and SmCoAsO well scale to each other with the difference of de Gennes factors of Nd^{3+} and Sm^{3+}. On the contrary, in the case of Ln = Gd, some different situation may occur compared with Nd and Sm cases. We draw the total magnetic phase diagram of LnCoAsO (Ln = La -Gd) from the results of magnetic measurements.

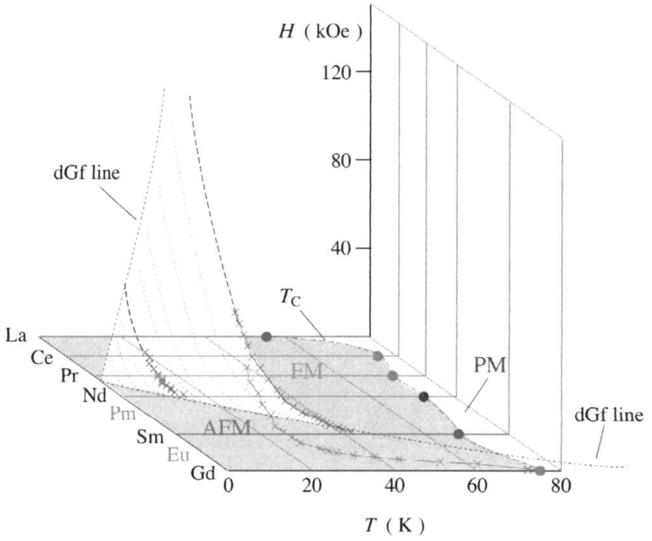

Figure 5. Total magnetic phase diagram of LnCoAsO (Ln = La - Gd). Dashed lines are guides for eyes. PM, FM, and AFM denote a paramagnetic phase, a ferromagnetic phase and an antiferromagnetic phase, respectively.

4.1. Acknowledgments

This workshop was supported in part by the Grant-in-Aid for the Global COE Program "The Next Generation of Physics, Spun from Universality and Emergence" from the Ministry of Education, Culture, Sports, Science and Technology (MEXT) of Japan. The authors thank the Yukawa Institute for Theoretical Physics at Kyoto University, where this work was initiated [completed] during the YITP-W-10-12 on "International and Interdisciplinary Workshop on Novel Phenomena in Integrated Complex Sciences: from Non-living to Living Systems". This work is supported by Grants-in-Aid for Scientific Research from the Japan Society for Promotion of Science (Grants No.19350030 and No.22350029) and also by Global COE Program "the Physical Sciences Frontier", MEXT, Japan.

References

[1] Kamihara Y, Watanabe T, Hirano M and Hosono H 2008 *J. Am. Chem. Soc.* **130** 3296–3297
[2] Ren Z A, Lu W, Yang J, Yi W, Chen X L, Li Z C, Che G C, Dong X L, Sun L L, Zhou F and Zhao Z X 2008 *Chin. Phys. Lett.* **25** 2215–2216
[3] Rotter M, Tegel M and Johrendt D 2008 *Phys. Rev. Lett.* **101** 107006
[4] Wang X C, Liu Q Q, Lv Y X, Gao W B, Yang L X, Yu R C, Li F Y and Jin C Q 2008 *Solid State Commun.* **148** 538–540
[5] Ogino H, Matsumura Y, Katsura Y, Ushiyama K, Horii S, Kishio K and ichi Shimoyama J 2009 *Supercond. Sci. Technol.* **22** 075008
[6] Hsu F C, Luo J Y, Yeh K W, Chen T K, Huang T W, Wu P M, Lee Y C, Huang Y L, Chu Y Y, Yan D C and Wu M K 2008 *Proc. Natl. Acad. Sci. U. S. A.* **105** 14262
[7] Yanagi H, Kawamura R, Kamiya T, Kamihara Y, Hirano M, Nakamura T, Osawa H and Hosono H 2008 *Phys. Rev. B* **77** 224431
[8] Sefat A S, Huq A, McGuire M A, Jin R, Sales B C, Mandrus D, Cranswick L M D, Stephens P W and Stone K H 2008 *Phys. Rev. B* **78** 104505
[9] Ohta H and Yoshimura K 2009 *Phys. Rev. B* **79** 184407
[10] Zimmer B I, Jeitschko W, Albering J H, Glaum R and Reehuis M 1995 *J. Alloys Comp.* **229** 238
[11] Quebe P, Terbüchte L T and Jeischko W 2000 *J. Alloys Comp.* **302** 70
[12] Ohta H and Yoshimura K 2009 *Phys. Rev. B* **80** 184409

[13] McGuire M A, Gout D J, Garlea V O, Sefat A S, Sales B C and Mandrus D 2010 *Phys. Rev. B* **81** 104405
[14] Marcinkova A, Grist D A M, Margiolaki I, Hansen T C, Margadonna S and Bos J W G 2010 *Phys. Rev. B* **81** 064511
[15] Awana V P S, Nowik I, Pal A, Yamaura K, Takayama-Muromachi E and Felner I 2010 *Phys. Rev. B* **81** 212501
[16] Ohta H, Michioka C, Matsuo A, Kindo K and Yoshimura K 2010 *Phys. Rev. B* **82** 054421

Journal of Physics: Conference Series **344** (2012) 012026 doi:10.1088/1742-6596/344/1/012026

^{77}Se-NMR study of quasi-one dimensional organic conductor $(TMTSF)_2X$

K. Nishiyama[a], T. Mito[a], Y. Kujirai[a], T. Koyama[a], K. Ueda[a], T. Kohara[a], K. Takeuchi[a], H. Akutsu[a], J. Yamada[a], A. Kornilov[b], V. M. Pudalov[b], and J. S. Qualls[c]

[a]Graduate school of Material Science, University of Hyogo, Hyogo, 678-1297, Japan
[b]P. N. Lebedev Physics Institute, Moscow 119991, Russia
[c]Sonoma State University, Rohnert Park, California 94928, USA

E-mail: mito@sci.u-hyogo.ac.jp

Abstract. Quasi-one dimensional organic conductor $(TMTSF)_2X$ with octahedral anions X = PF_6, AsF_6, and SbF_6 have been studied by NMR measurements at the Se-sites which are substantially responsible for conductivity in this system. Close resemblances in the temperature dependence of the spin-lattice relaxation rate $1/T_1$ and the evolution of internal field in the SDW state were observed in the three measured compounds. The existence of the sub-phases in the SDW phase of all these compounds, which is evidenced by T_1 but not from spectrum measurements, indicates that this is a common feature among $(TMTSF)_2X$ system with an incommensurate SDW wave number and is associated with dynamic property rather than static one.

1. Introduction

The first organic superconductor $(TMTSF)_2X$ (X= PF_6, AsF_6, SbF_6, ClO_4, etc.) has been a topic of interest for three decades. Intensive studies have been carried out on this family, however, there still remain several basic problems to be clarified, including nature of the unconventional superconducting phase adjacent to the spin density wave (SDW) ordered state [1]. Besides, the existence of sub-phases in the SDW phase was pointed out by ^{1}H-NMR studies [2, 3], whose mechanism remains unsolved as well. Nuclear magnetic resonance (NMR) is a powerful experimental technique to probe static and dynamic magnetic properties. In contrast to detailed ^{1}H-NMR studies on these compounds, there are only few ^{77}Se-NMR reports using single crystals in the SDW state [4], although Se-4s and -4p orbitals mainly contribute to the conduction bands [5]. We have therefore carried out ^{77}Se-NMR measurements on single crystal samples of $(TMTSF)_2X$ with octahedral anions X = PF_6, AsF_6, and SbF_6. Such a systematic study over several isostructural analogues will enable one to conduct comprehensive investigations on this system. The SDW phase of X= PF_6 is characterized by an incommensurate wave vector of $\boldsymbol{Q}_{\rm SDW} = (0.5, 0.24 \pm 0.03, -0.06 \pm 0.20)$ from ^{1}H-NMR study [6], and SDW transition temperature $T_{\rm SDW}$ is 12 K for X= PF_6 and AsF_6. Regarding X= SbF_6, while $T_{\rm SDW}$ was first reported as 17 K [7], another reference indicates $T_{\rm SDW}$ = 12 K [8].

2. Experimental details

Single crystals were grown by the standard electro-chemical method. The ^{77}Se-NMR measurements were carried out using the spin echo technique with a phase-coherent pulsed spectrometer. NMR lines were obtained by sweeping magnetic field, and the spin lattice relaxation time T_1 was measured with a single rf-pulse saturation method. All the NMR measurements were performed around the magnetic field of 11 – 12 T using a commercial superconducting magnet, with magnetic field direction parallel to the crystal c^*-axis.

3. Results and Discussion

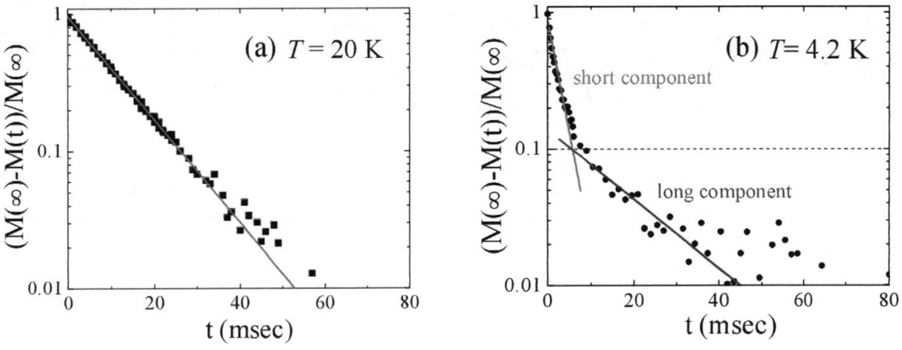

Figure 1. Representative T_1 recovery curves (a) above and (b) below $T_{\rm SDW}$ for present $(\mathrm{TMTSF})_2X$ samples. The presented data were obtained for $X = \mathrm{SbF_6}$ at the frequency of 92.680 MHz corresponding to the field of $\sim$ 11.4 T. While T_1 was uniquely determined in the paramagnetic state, we evaluated T_1 with short and long components below $T_{\rm SDW}$ (see text). The short and long components were temporally obtained by fitting the data of $0.1 \leq (M(\infty) - M(t))/M(\infty) \leq 1$ and $(M(\infty) - M(t))/M(\infty) \leq 0.1$ to the single exponential function, respectively. Here, $M(\infty)$ and $M(t)$ are the nuclear magnetization in the thermal equilibrium and at a time t after a saturation pulse, respectively.

Figure 1 shows representative T_1 recovery curves obtained above and below $T_{\rm SDW}$ for $X = \mathrm{SbF_6}$. Since ^{77}Se nucleus has the nuclear spin $I = 1/2$, the recovery curves in the paramagnetic state were fit well to a single exponential function. On the other hand, we could not obtain unique T_1 below $T_{\rm SDW}$ for all the three compounds, possibly due to distribution of internal field in the SDW state with incommensurate $\boldsymbol{Q}$-vector. Hence we evaluated the T_1 data with short and long T_1 components, as shown in Fig. 1(b).

The temperature dependences of $1/T_1$ for the three compounds are summarized in Fig. 2(a)-(c). At high temperatures above 30 K, $1/T_1$ shows rather metallic behavior, namely the Korringa law $1/T_1 \propto T$. At low temperatures, the temperature dependence of $1/T_1$ is analogous with the previous ^{77}Se-NMR study [4]. $1/T_1$ shows critical slowing down behavior when approaching $T_{\rm SDW}$ from high temperatures. Additionally $1/T_1$ shows a shoulder-like anomaly at $T_X \sim 4$ K in the SDW state. Note that, in the temperature region from T_X to $T_{\rm SDW}$, $1/T_1$ follows rather $T^{1.5}$ than T indicated in Ref. [4]. The difference is possibly associated with the different resonance line shape in the SDW state between our results and that in Ref. [4]. Namely, in the present study, the NMR line significantly broadens and splits into double-peaked lines below $T_{\rm SDW}$, as shown in Fig. 3(b) and (c) (also see Ref. [9]), so that the T_1 measurements were performed at

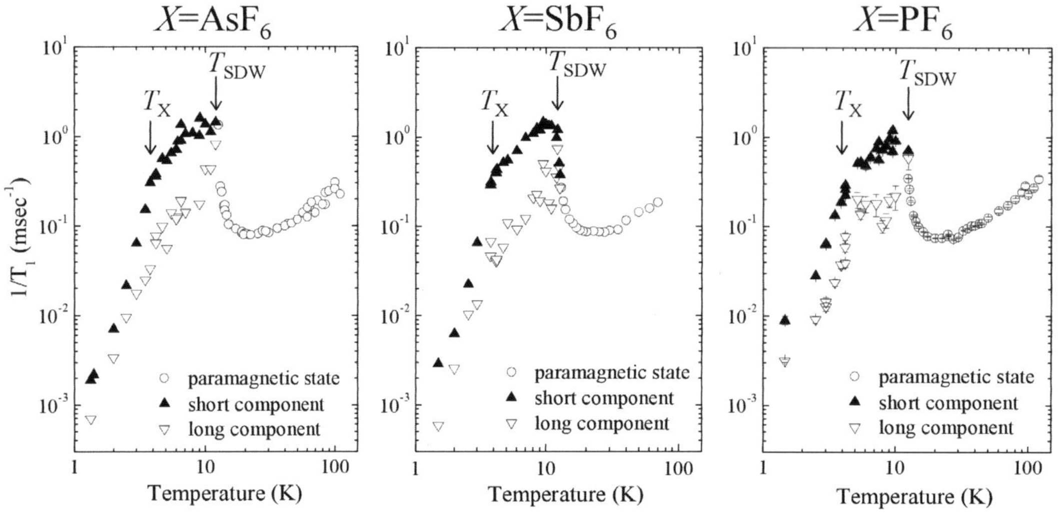

Figure 2. Summary of the temperature dependences of $1/T_1$ for X = AsF_6, SbF_6, and PF_6 compounds. For all these compounds, T_1 in the SDW state was evaluated with short and long components as shown in Fig. 1.

the peaks of the split lines. Here we did not see any significant discrepancy between the T_1 data measured at the two different peaks. On the other hand, no clear splitting is seen in the NMR spectrum reported in Ref. [4]. The double-peak line shape is reproduced by assuming sinusoidal internal fields at Se nucleus positions having the same incommensurate wave number with the SDW $\boldsymbol{Q}$-vector [9]. The line width at 4.2 K corresponds to $\sim$ 800 Oe, larger than the value previously reported [4]. Thus the T_1 relaxation process in this anomalous state between T_X and $T_{\rm SDW}$ is not only attributed to residual free carriers in the SDW state, but also to magnetic mechanism.

NMR spectrum measurement is a useful tool to investigate the evolution of the static SDW state. In Fig. 3(a), we plot the normalized interval of the split NMR lines, $\Delta(T)/\Delta_0$, observed in the SDW state as a function of temperature. Δ_0 was obtained by extrapolating $\Delta(T)$ as $T \to 0$. From this plot, onset temperature for the SDW order is determined to be $12.0-12.5$ K for all these compounds. The evolution of the internal field is steeper than the BCS curve just below $T_{\rm SDW}$, consistent with the report of ^{1}H-NMR [2]. Contrary to the remarkable anomaly in $1/T_1$ at T_X, NMR line does not show any change in the line shape within the experimental accuracy, indicating that the transfer between the sub-phases at T_X is mainly brought about by dynamic properties [9].

Apparently our results show that the anomaly at the sub-phase boundary $T_X \sim 4$ K observed in X= PF_6 (^{1}H- and ^{77}Se-NMR) [2, 3, 4] and AsF_6 (^{1}H-NMR) [10] is also seen in X= SbF_6 compound. Because, for $(TMTTF)_2Br$, a corresponding anomaly is not detected in the commensurate SDW state [3], our findings support the view that the existence of the sub-phases in the SDW phase is a common feature among $(TMTSF)_2X$ system with an incommensurate SDW wave number. As shown in Table 1, the lattice parameters, as well as the unit cell volume, of X= PF_6 and AsF_6 are close to each other. On the other hand, X= SbF_6 has $2-3$ % larger volume than others, which brings an expectation of negative pressure effects on physical properties. The observed close resemblances in $1/T_1$ and $\Delta(T)/\Delta_0$ between X= SbF_6 and others are probably consistent with the regime that the ratio of the strongest electron transfer energy along the $a-$axis and the perturbational transfer along $b'-$direction is responsible for

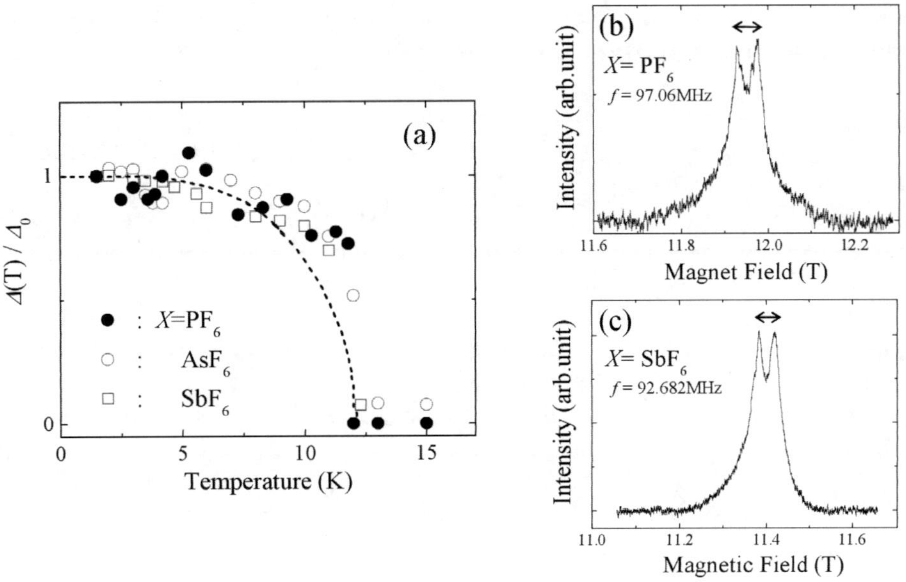

Figure 3. (a) The normalized interval of the split lines $\Delta(T)/\Delta_0$ as a function of temperature for X = AsF_6, SbF_6, and PF_6 compounds. The dotted line shows a BCS curve. (b) and (c) show representative lines for X = PF_6 and SbF_6, respectively. Both were measured at 1.5 K and at the NMR frequency f. The arrows represent $\Delta(T)$.

determination of T_{SDW}, and the smallest transfer in the c−direction is negligible, as indicated in the Ref. [11].

Table 1. Lattice parameters of X = PF_6, AsF_6 from Ref. [7] and of the present X = SbF_6 sample. V is the volume of the triclinic unit cell. The data for the present X = SbF_6 sample is in good agreement with those from Ref. [7].

X	a (Å)	b (Å)	c (Å)	V (Å^3)
PF_6	7.297	7.710	13.522	714.0
AsF_6	7.277	7.711	13.651	719.9
SbF_6	7.294	7.726	13.908	737.7

4. Conclusion

Systematic ^{77}Se-NMR study have been carried out on single crystal $(TMTSF)_2X$ with octahedral anions X = PF_6, AsF_6, and SbF_6. The anomaly at the sub-phase boundary $T_X \sim 4$ K, which were previously indicated by ^{1}H- and ^{77}Se-NMR on X= PF_6 [2, 3, 4] and ^{1}H-NMR on X= AsF_6 [10], was almost equally observed in the temperature dependence of $1/T_1$ but not in the spectrum measurements on the three measured compounds. Our results obviously shows that the existence of the sub-phases is common over the $(TMTSF)_2X$ with an incommensurate SDW wave number, and it originates from dynamic features rather than static ones. It is interesting that, in spite of $2-3$ % difference in the unit cell volume, close resemblances in the temperature dependence of $1/T_1$ and in the evolution of internal field in the SDW state were observed in all these compounds. More detail studies, especially NMR measurements with different angle external fields and/or under pressure, are in progress.

Acknowledgment
This workshop was supported in part by the Grant-in-Aid for the Global COE Program "The Next Generation of Physics, Spun from Universality and Emergence" from the Ministry of Education, Culture, Sports, Science and Technology (MEXT) of Japan.

The authors thank Y. Hasegawa for valuable discussions. We also thank the Yukawa Institute for Theoretical Physics at Kyoto University. Discussions during the YITP workshop YITP-W-10-12 on "International and Interdisciplinary Workshop on Novel Phenomena in Integrated Complex Sciences: from Non-living to Living Systems" were useful to complete this work. This work was partially supported by Grant-in-Aid for Scientific Research (Grant No. 21540342) from the Ministry of Education, Culture, Sports, Science and Technology of Japan.

References

[1] Jerome D, Mazaud A, Ribault M and Bechgaard K 1991 J. Physique Lett. **41** 95
[2] Takahashi T, Harada T, Kobayashi Y, Kanoda K, Suzuki K, Murata K and Saito G 1991 Synth. Met. **41-43** 3985
[3] Nomura K, Kubota H, Kotomizu M, Hanajiri T, Nakatsuji S and Yamada J 2003 Synth. Met. **133-134** 19
[4] Valfells S, Kuhns P, Kleinhammes A, Brooks J S, Moulton W, Takasaki S, Yamada J and Anzai H 1997 Phys. Rev. B **56** 2585
[5] Azevedo L J, Schirber J E and Engler E M 1984 Phys. Rev. B. **29** 464
[6] Takahashi T, Maniwa Y, Kawamura H and Saito G 1986 J. Phys. Soc. Jpn. **55** 1364
[7] Parkin S S P, Ribault M, Jerome D and Bechgaard K 1981 J. Phys. C: Solid State Phys. **14** 5305
[8] Vignolles D, Ulmet J P, Naughton M J, Binet L and Fabre J M 1998 Phys. Rev. B **58** 14 476
[9] Mito T, Nishiyama K, Koyama T, Ueda K, Kohara T, Takeuchi K, Akutsu H, Yamada J, Kornilov A, Pudalov V M and Qualls J S 2010 Physica C **470** S592
[10] Nomura K, Hosokawa Y, Matsunaga N, Nagasawa M, Sambongi T and Anzai H 1995 Synth. Met. **70** 1295
[11] Yamaji K 1982 J. Phys. Soc. Jpn. **51** 2787

Journal of Physics: Conference Series **344** (2012) 012027 doi:10.1088/1742-6596/344/1/012027

Nuclear Quadrupole Resonance in the Heavy Fermion Antiferromagnet $CePt_2In_7$

N apRoberts-Warren, A P Dioguardi, A C Shockley, C H Lin, J Crocker, P Klavins and N J Curro

Department of Physics, University of California, Davis, CA 95616 USA

naprobertswarren@ucdavis.edu

New ^{115}In NQR data is presented for the 9/2 ↔ 7/2 of the In(2) site transition in the heavy fermion compound $CePt_2In_7$. We extract the sub-lattice magnetization in the antiferromagnetic state. The spectra reveal that roughly half of the In(2) sites experience no static hyperfine field.

1. Introduction

In the conventional BCS theory of superconductivity, magnetic atoms tend to destabilize the superconducting ground state. It is therefore surprising that several new classes of superconductors, including the Heavy Fermion (HF) and High Termperature Superconducting Cuprate (HTSC) materials, exhibit phase diagrams with adjacent superconducting and antiferromagnetic (AFM) regimes. The discovery of superconductivity under pressure in cubic HF $CeIn_3$ [1,2] and the enhanced T_C's in the tetragonal $CeMIn_5$ class hints at the important role of reduced dimensionality for unconventional superconductivity. In the $CeMIn_5$ family, MIn_2 planar squares are interleaved between the Ce-In planes. This expands the bond length c/a ratio to ~1.6, reducing the dimensionality and enhancing T_C by an order of magnitude. The so-called 115 materials exhibit a rich spectrum of behavior, from ambient pressure superconductivity in $CeCoIn_5$ [4] to incommensurate long-range AFM ordering in $CeRhIn_5$ [5]. Furthermore, the phase diagrams for both $CeIn_3$ and $CeRhIn_5$ reveal co-existence of superconductivity and magnetic order [6]. This proximity of superconductivity to antiferromagnetism in anisotropic crystal structures continues to attract significant attention.

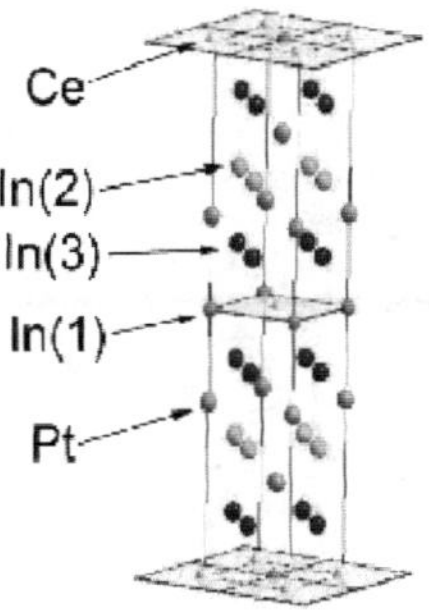

Figure 1. the unit cell of $CePt_2In_7$

$CePt_2In_7$ is another member of the Ce-In heavy fermion family [7,8]. It crystallizes in the *I*4/*mmm* space group with unit cell parameters a = 4.602 Å and c = 21.601 Å, increasing the spacing between the Ce-In planes by 43% from $CeRhIn_5$. This material exhibits has the classic signature of a Heavy Fermion: an enhanced Sommerfeld coefficient of 450 mJ/mol∘K at low temperature, and an effective moment $\mu = 2.41\mu_B$ consistent with Ce^{3+}. At ambient pressure, $CePt_2In_7$ undergoes a paramagnetic to AFM phase transition at T_N = 5.2 K, and resistance experiments under pressure revealed clear evidence of a superconducting transition of T_C = 1 K at 2.3 GPa [8]. In this paper we present NQR data on the sub-lattice magnetization at ambient pressure.

2. Experimental Procedure

High quality crystalline samples were obtained by arc-melting the atoms Ce, Pt, and In together in the ratio 1:2:7. The resulting pellet was then wrapped in Ta foil and sealed under vacuum in a quartz ampoule. The ampoule was then placed in an oven and annealed for several days. Upon cracking the Ta envelope, regular crystallite formation was observed. The $CePt_2In_7$ crystallites were subjected to powder x-ray diffraction in a Siemens Diffraktometer to verify the crystal structure, and heat capacity and resistance measurements confirmed the onset of magnetic order at T_N = 5.2 K. No impurity peaks (such as the AFM transition for $CeIn_3$ at 10.2 K) were observed in the transport measurements, again verifying the high-quality of the sample (x-ray diffraction identified < 3% impurity phase).

3. ^{115}In NQR

$CePt_2In_7$ is ideal for NQR measurements because there are three distinct crystallographic In sites. NQR spectra were obtained at the the 9.5 MHz transition of the In(2) site between 6 K and 1.5 K. Spin echoes were obtained as function of frequency, and the echo integral is plotted in Figure 3c.

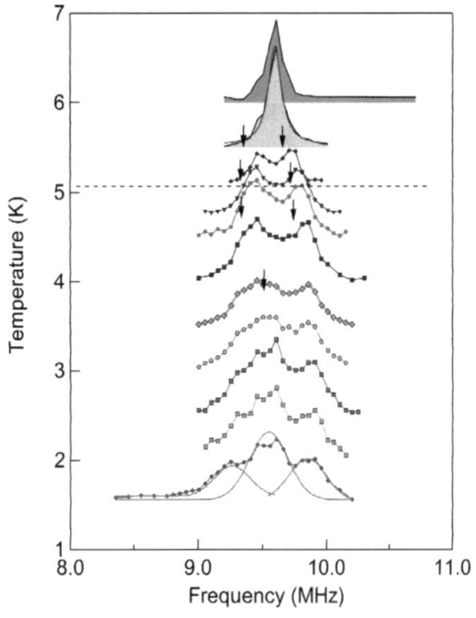

Figure 2. Evolution of In(2) NQR resonance through the AFM transition with gaussian fits (solid red lines) used to assess spectral weight attached to the 1.5 K spectra

The nuclear spin Hamiltonian is

$$H = \gamma h \vec{H}_{int} \cdot \vec{I} + h\nu_Q/6(3I_Z^2 - I^2 + \eta(I_X^2 - I_Y^2)) \quad ,$$

where $h\nu_Q = 3eQV_{ZZ}/2I(2I\text{-}1)$ is the quadrupolar splitting frequency, V_{ZZ} is the electric field gradient (EFG) along the principal axis and η is the axial asymmetry parameter and $\overline{H}_{int}$ is the static internal hyperfine field from the ordered Ce moments. For In(2), ν_Q = 2.5 MHz and η = 0.4. As seen in Figure 2, the $4\nu_Q$ resonance is split by the presence of an internal field. The splitting looks relatively well-defined from T_N down to 4 K; below this significant spectral weight develops again at 9.6 MHz, in addition to the two split peaks. The upper-frequency (right-hand) remains discernible all the way to 1.7 K, and provides an estimate of internal field at the In(2) site below T_N (see Figure 3a). It was then fit to the function

$$H_{int}(T) = H_{int}^0(1 - (T/T_N))^\beta \quad ,$$

with values H^0_{int} = 332.46 Oe and β = 0.22, lower then the 3D mean-field result β = 0.5 (Figure 3a) [5].

4. Magnetic Structure

Several key pieces of information can be drawn from these spectra. The fact that the splitting is symmetric about the original resonance line indicates that the internal field lies parallel to the principal EFG axis, $\bar{q}$. A priori we do not know the direction of $\bar{q}$, however in the $CeMIn_5$ materials $\bar{q}$ lies in the ab plane and points towards the center of the unit cell at the In(2) sites. We assume that $\bar{q}$, and hence the internal field at the In(2) sites in $CePt_2In_7$, points in the ab plane. However, without a single crystal to examine the angular field dependence of the resonance line, we conclusively cannot determine the direction of $\bar{q}$. It is important to point out, however, that the In(2) site in the 115's is analogous to the In(3), not In(2), site in $CePt_2In_7$.

Surprisingly at ~4 K, a new feature emerges between the two horns of the split resonance line. The spectral weight of this feature increases with decreasing temperature and reaches 40% by 1.5 K (see Fig 3c). The origin of this three-peaked structure remains unclear, but suggests a change in the magnetic ordering wave-vector. Such a transition has been observed before in the iron-arsenides [9] but no evidence of any phase change is seen in the transport property or specific heat measurements in $CePt_2In_7$.

In conclusion, we have measured the sublattice magnetization in $CePt_2In_7$ below T_N. The evolution of magnetic order at In(2) suggests commensurate magnetic order, yet the anomalous re-appearence of signal in the center remains unexplained. In order to determine the magnetic structure in this crystal, high-quality single crystals are necessary, which will enable angular field-dependent studies to ascertain the principal axes of the EFG and will also enable direct neutron scattering measurements.

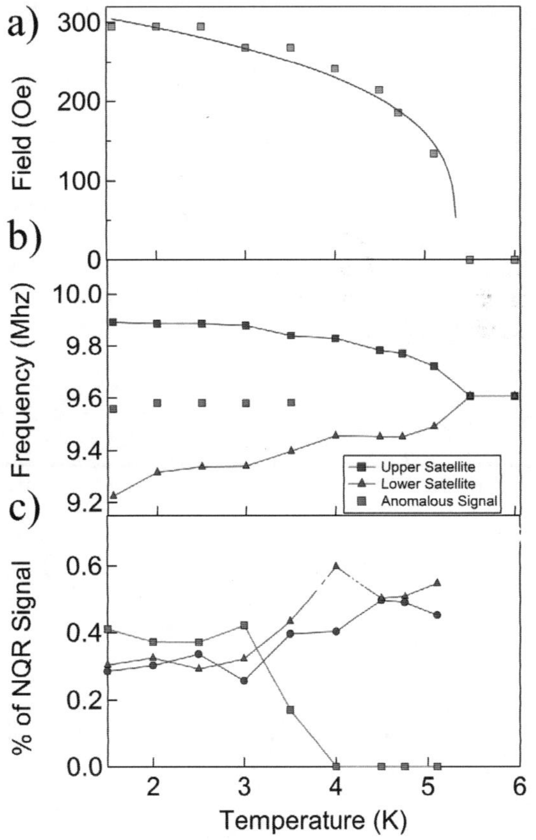

Figure 3. a) internal field at In(2) site below T_N, b) the NQR frequencies of the AFM satellites splitting and the appearance of the central signal, and c) the normalized area of the AFM satellites and anomalous signal

5. Acknowledgements

The authors thank the Yukawa Institute for Theoretical Physics at Kyoto University. Discussions during the YITP workshop YITP-W-10-12 on "International and Interdisciplinary Workshop on Novel Phenomena in Integrated Complex Sciences: from Non-living to Living Systems" were useful to complete this work. The authors would also like to thank the conference organizers and ICAM/I2CAM for travel funding. We thank D. Pines for enlightening discussions.

References

[1] Knebel G, Braithwaite D, Canfield P, Lapertot G, and Flouquet J, *Phys. Rev. B* **65**(2), 024425 (2001)
[2] Mathur N, Grosche F, Julian S, Walker I, Freye D, Haselwimmer R, and Lonzarich G, *Nature* **394**, 39 (1998)
[3] Hegger H, Petrovic C, Moshopoulou E, Hundley M, Sarrao J, Fisk Z, and Thompson J, *Phys. Rev. Lett.* **84**, 4986 (2000)
[4] Petrovic C, Movshovich R, Jaime M, Pagliuso P, Hundley M, Sarrao J, Fisk Z, and Thompson J, Europhys. Lett. **53**, 354 (2001)
[5] Curro N, Hammel P, Pagliuso P, Sarrao J, Thompson J, and Fisk Z, *Phys. Rev. B* **62**(10), 6100-6103(R) (2000)
[6] Sebastian S, Harrison N, Batista C, Trugman S, Fanelli V, Jaime M, Murphy T, Palm E, Harima H, and Ebihara T, *Proc Nat'l Acad Sci USA*. 2009 May 12; **106**(19): 7741–44
[7] apRoberts-Warren N, Dioguardi A, Shockley A, Lin C, Crocker J, Klavins P, and Curro N, *Phys. Rev. B* **81**, 180403(R) (2010)
[8] Bauer E, Sidorov V, Lee H, Kurita N, Ronning F, Movshovich R, and Thompson J, *J. Phys.: Conf. Ser.* **200,** 012011 (2010)
[9] Kitagawa K, Mezaki Y, Matsubayashi K, Uwatoko Y, and Takigawa M, *arxiv/cond-mat*:1011.1108v1

AUTHOR INDEX

Adroja, D T....1
Akutsu, H....160
apRoberts-Warren, N....129, 139, 165
Bud'Ko, S L....139
Canfield, P C....139
Chen, B....125
Crocker, J....165
Curro, N J....129, 139, 165
Dioguardi, A P....129, 139, 165
Ge, W....143
Gold, C....1
Hatayama, N....100, 107
Hillier, A D....1
Hirota, K....77, 113
Hondou, T....39, 48
Hongo, S....77
Imai, M....88
Inagawa, M....27
Irizawa, A....96
Ishido, M....39, 48
Ishizawa, K-I....39, 48
Kambe, S....19
Kaneko, K....19
Kato, M....77, 113
Kindo, K....148
Kitazawa, H....88
Klavins, P....129, 165
Kohara, T....160
Konno, R....100, 107
Kornilov, A....160
Kotani, Y....117
Koyama, T....160
Koza, M M....1
Kujirai, Y....160
Kunita, I....33
Lin, C H....165
Makino, Y....133
Makoshi, K....117
Manalo, S....1
Matsuo, A....148
Michioka, C....125, 143, 148, 153
Michor, H....1
Mito, T....160
Miyata, H....39, 48
Murase, M....39, 48
Murase, Y....113
Naitoh, K....53
Nakamura, H....83, 113
Nanba, T....96
Nawa, K....148
Ni, N....139
Nishimura, K....27
Nishiyama, K....160
Nogi, Y....77
Ohta, H....125, 143, 153
Okada, A....27
Otsuka, Y....71
Peyker, L....1
Pudalov, V M....160
Qualls, J S....160
Sakai, H....19
Sakai, T....71
Sakurai, H....96
Sakurazawa, S....33
Scheidt, E -W....1
Shima, N....117
Shockley, A C....129, 139, 165
Sugawara, K....39
Tabata, Y....83, 113
Takahashi, Y....9, 100
Takehara, S....125
Takeuchi, K....160
Takigawa, M....83
Terazawa, S....83, 113
Tobinaga, Y....27
Tokunaga, Y....19
Tsujii, N....88
Ueda, K....160
Umemoto, Y....113
Waki, T....83, 113
Yamada, J....160
Yang, J....125
Yoshimura, K....125, 133, 143, 148, 153